Case Studies in Forensic Physics

Synthesis Lectures on Engineering, Science, and Technology

Each book in the series is written by a well known expert in the field. Most titles cover subjects such as professional development, education, and study skills, as well as basic introductory undergraduate material and other topics appropriate for a broader and less technical audience. In addition, the series includes several titles written on very specific topics not covered elsewhere in the Synthesis Digital Library.

Case Studies in Forensic Physics
Gregory A. DiLisi and Richard A. Rarick
2020

Integrated Process Design and Operational Optimization via Multiparametric Programming
Baris Burnak, Nikolaos A. Diangelakis, and Efstratios N. Pistikopoulos
2020

Nanotechnology Past and Present
Deb Newberry
2020

Introduction to Engineering Research
Wendy C. Crone
2020

Theory of Electromagnetic Beams
John Lekner
2020

The Search for the Absolute: How Magic Became Science
Jeffrey H. Williams
2020

The Big Picture: The Universe in Five S.T.E.P.S.
John Beaver
2020

Relativistic Classical Mechanics and Electrodynamics
Martin Land and Lawrence P. Horwitz
2019

Generating Functions in Engineering and the Applied Sciences
Rajan Chattamvelli and Ramalingam Shanmugam
2019

Transformative Teaching: A Collection of Stories of Engineering Faculty's Pedagogical Journeys
Nadia Kellam, Brooke Coley, and Audrey Boklage
2019

Ancient Hindu Science: Its Transmission and Impact on World Cultures
Alok Kumar
2019

Value Rational Engineering
Shuichi Fukuda
2018

Strategic Cost Fundamentals: for Designers, Engineers, Technologists, Estimators, Project Managers, and Financial Analysts
Robert C. Creese
2018

Concise Introduction to Cement Chemistry and Manufacturing
Tadele Assefa Aragaw
2018

Data Mining and Market Intelligence: Implications for Decision Making
Mustapha Akinkunmi
2018

Empowering Professional Teaching in Engineering: Sustaining the Scholarship of Teaching
John Heywood
2018

The Human Side of Engineering
John Heywood
2017

Geometric Programming for Design Equation Development and Cost/Profit
Optimization (with illustrative case study problems and solutions), Third Edition
Robert C. Creese
2016

Engineering Principles in Everyday Life for Non-Engineers
Saeed Benjamin Niku
2016

A, B, See... in 3D: A Workbook to Improve 3-D Visualization Skills
Dan G. Dimitriu
2015

The Captains of Energy: Systems Dynamics from an Energy Perspective
Vincent C. Prantil and Timothy Decker
2015

Lying by Approximation: The Truth about Finite Element Analysis
Vincent C. Prantil, Christopher Papadopoulos, and Paul D. Gessler
2013

Simplified Models for Assessing Heat and Mass Transfer in Evaporative Towers
Alessandra De Angelis, Onorio Saro, Giulio Lorenzini, Stefano D'Elia, and Marco Medici
2013

The Engineering Design Challenge: A Creative Process
Charles W. Dolan
2013

The Making of Green Engineers: Sustainable Development and the Hybrid Imagination
Andrew Jamison
2013

Crafting Your Research Future: A Guide to Successful Master's and Ph.D. Degrees in
Science & Engineering
Charles X. Ling and Qiang Yang
2012

Fundamentals of Engineering Economics and Decision Analysis
David L. Whitman and Ronald E. Terry
2012

A Little Book on Teaching: A Beginner's Guide for Educators of Engineering and
Applied Science
Steven F. Barrett
2012

Engineering Thermodynamics and 21st Century Energy Problems: A Textbook Companion for Student Engagement
Donna Riley
2011

MATLAB for Engineering and the Life Sciences
Joseph V. Tranquillo
2011

Systems Engineering: Building Successful Systems
Howard Eisner
2011

Fin Shape Thermal Optimization Using Bejan's Constructal Theory
Giulio Lorenzini, Simone Moretti, and Alessandra Conti
2011

Geometric Programming for Design and Cost Optimization (with illustrative case study problems and solutions), Second Edition
Robert C. Creese
2010

Survive and Thrive: A Guide for Untenured Faculty
Wendy C. Crone
2010

Geometric Programming for Design and Cost Optimization (with Illustrative Case Study Problems and Solutions)
Robert C. Creese
2009

Style and Ethics of Communication in Science and Engineering
Jay D. Humphrey and Jeffrey W. Holmes
2008

Introduction to Engineering: A Starter's Guide with Hands-On Analog Multimedia Explorations
Lina J. Karam and Naji Mounsef
2008

Introduction to Engineering: A Starter's Guide with Hands-On Digital Multimedia and Robotics Explorations
Lina J. Karam and Naji Mounsef
2008

CAD/CAM of Sculptured Surfaces on Multi-Axis NC Machine: The DG/K-Based Approach
Stephen P. Radzevich
2008

Tensor Properties of Solids, Part Two: Transport Properties of Solids
Richard F. Tinder
2007

Tensor Properties of Solids, Part One: Equilibrium Tensor Properties of Solids
Richard F. Tinder
2007

Essentials of Applied Mathematics for Scientists and Engineers
Robert G. Watts
2007

Project Management for Engineering Design
Charles Lessard and Joseph Lessard
2007

Relativistic Flight Mechanics and Space Travel
Richard F. Tinder
2006

Case Studies in Forensic Physics
Gregory A. DiLisi and Richard A. Rarick

ISBN: 978-3-031-00958-7   paperback
ISBN: 978-3-031-02086-5   ebook
ISBN: 978-3-031-00158-1   hardcover

DOI 10.1007/978-3-031-02086-5

A Publication in the Springer series
SYNTHESIS LECTURES ON ENGINEERING, SCIENCE, AND TECHNOLOGY

Lecture #9
Series ISSN
Print 2690-0300   Electronic 2690-0327

Willy Stöwer, *Der Untergang der Titanic*, 1912. Public domain.

Case Studies in Forensic Physics

Gregory A. DiLisi
John Carroll University, Univesity Heights, Ohio

Richard A. Rarick
Cleveland State University, Cleveland, Ohio

SYNTHESIS LECTURES ON ENGINEERING, SCIENCE, AND TECHNOLOGY #9

ABSTRACT

This book focuses on a forensics-style re-examination of several historical events. The purpose of these studies is to afford readers the opportunity to apply basic principles of physics to unsolved mysteries and controversial events in order to settle the historical debate. We identify nine advantages of using case studies as a pedagogical approach to understanding forensic physics. Each of these nine advantages is the focus of a chapter of this book. Within each chapter, we show how a cascade of unlikely events resulted in an unpredictable catastrophe and use introductory-level physics to analyze the outcome. Armed with the tools of a good forensic physicist, the reader will realize that the historical record is far from being a set of agreed upon immutable facts; instead, it is a living, changing thing that is open to re-visitation, re-examination, and re-interpretation.

KEYWORDS

forensic physics, applied physics, forensic analysis, introductory physics

This book is dedicated to my family:
to my grandparents, Tommaso and Carmela Frate,
to my parents, Richard and Mary DiLisi,
to my siblings, Rick DiLisi, Carla Solomon, and Jennifer Newton,
to my wife, Linda,
to my daughter, Carmela,
and
to the wonderful creatures who inhabit our home.

Gregory A. DiLisi

Contents

Preface xix

Acknowledgments xxv

1   Taking a Forensics Approach to History 1

1.1   Bouncing Back from "Deflategate": A Case Study in the Physics of a
Bouncing Ball 1

Gregory A. DiLisi
John Carroll University, University Heights, Ohio

Richard A. Rarick
Cleveland State University, Cleveland, Ohio

1.1.1   Introduction 1
1.1.2   The Controversy 3
1.1.3   The Media Blitz–Physics to the Rescue 6
1.1.4   A New Focus for "Deflategate" 8
1.1.5   Statement of the Problem 8
1.1.6   The Physics of a Bouncing Ball 9
1.1.7   Phases of a Bouncing Ball and the Coefficient of Restitution 10
1.1.8   Experimental Results and Discussion 13
1.2   Conclusions 25
1.3   References 25

2   Having Interdisciplinary Appeal 27

2.1   Holy High-Flying Hero! Bringing a Superhero Down to Earth: A Case
Study in Uniformly Accelerated Motion 27

Gregory A. DiLisi
John Carroll University, University Heights, Ohio

2.1.1   Introduction 27
2.1.2   It's a Bird… It's a Plane… It's the Hooded Llama! 29
2.1.3   Statement of the Problem 30
2.1.4   Meanwhile, Back in the Physics Laboratory… 32

2.2 Acknowledgments 35
2.3 References 35

3   Raising Historical Awareness and Bringing History to New Generations 37
3.1 The *Lady be Good*: A Case Study in Radio Frequency Direction Finders 38

Gregory A. DiLisi
John Carroll University, University Heights, Ohio

Alison Chaney
John Carroll University, University Heights, Ohio

Br. Kenneth Kane, C.S.C., KG8DN
Gilmour Academy, Gates Mills, Ohio

Robert L. Leskovec, K8DTS
RALTEC®div GENVAC Aerospace, Highlands Hts., Ohio

    3.1.1   Introduction 38
    3.1.2   The Mysterious Disappearance of the *Lady be Good* 39
    3.1.3   Radio Frequency Direction Finders 42
    3.1.4   Statement of the Problem 44
    3.1.5   Mystery Solved: *"The 180°-Ambiguity"* 47
3.2 Conclusions 53
3.3 Acknowledgments 54
3.4 References 54

4   Using Operational Definitions 57
4.1 The Hindenburg Disaster: Combining Physics and History in the
Laboratory, a Case Study in the Flammability of Fabrics (Vertical Flame
Tests) 57

Gregory A. DiLisi
John Carroll University, University Heights, Ohio

    4.1.1   Introduction 57
    4.1.2   "This Great Floating Palace" 58
    4.1.3   "The Ship is Riding Majestically Toward Us" 59
    4.1.4   "It's Burst into Flames" 61
    4.1.5   Theories 62
    4.1.6   Statement of the Problem 64
    4.1.7   Analysis 66
    4.1.8   Results 67

4.2   Conclusions 68

4.3   Acknowledgments 69

4.4   References 69

5   Demonstrating the Phenomenon of "Normalization of Deviance" 71

5.1   The Apollo I Fire: A Case Study in the Flammability of Fabrics
(Horizontal Flame Test) 72

Gregory A. DiLisi
John Carroll University, University Heights, Ohio

Stella McLean
John Carroll University, University Heights, Ohio

5.1.1   Introduction 72

5.1.2   The Crew 73

5.1.3   *"Go Fever!"* 74

5.1.4   *"We've Got a Fire in the Cockpit"* 75

5.1.5   Statement of the Problem 78

5.1.6   The Apollo Cabin 78

5.1.7   Sample Preparation 79

5.1.8   Testing and Results 79

5.1.9   *"The Kranz Dictum"* 82

5.2   Conclusions 83

5.3   References 83

6   Demonstrating "The Perfect Storm Scenario" 85

6.1   Remembering the S. S. Edmund Fitzgerald: A Case Study in Rogue Waves . 85

Gregory A. DiLisi
John Carroll University, University Heights, Ohio

Richard A. Rarick
Cleveland State University, Cleveland, Ohio

6.1.1   Introduction 85

6.1.2   The Mighty Fitz 86

6.1.3   The Final Voyage 88

6.1.4   The Wreckage Site 88

6.1.5   Theories 90

6.1.6   Rogue Waves 91

6.1.7   Statement of the Problem 92

      6.1.8   Results 95

  6.2   References 96

7 Developing Simulations and Testing Analogs and Proxies 97

  7.1   Modeling the 2004 Indian Ocean Tsunami for Introductory Physics
        Students: A Case Study in the Shallow Water Wave Equations 98

Gregory A. DiLisi

John Carroll University, University Heights, Ohio

Richard A. Rarick

Cleveland State University, Cleveland, Ohio

      7.1.1   Introduction 98

      7.1.2   Statement of the Problem 99

      7.1.3   Characteristics of a Tsunami 99

      7.1.4   The Indian Ocean and the 2004 Tsunami 101

      7.1.5   Simulation 1 – Using a *"Tsunami Tank"* 101

         7.1.5.1   Building a Tank 101

         7.1.5.2   Simulations 104

         7.1.5.3   Conclusions 106

      7.1.6   Simulation 2 – Computer-Based Model 108

         7.1.6.1   Step 1: The Shallow Water Wave Assumption 109

         7.1.6.2   Step 2: The Resulting Pressure Gradient 109

         7.1.6.3   Step 3: The Convective Derivative ("The Momentum
                 Equations") 110

         7.1.6.4   Step 4: "The Mass Continuity Equation" 111

      7.1.7   Computer Modeling 112

      7.1.8   Results 112

  7.2   Conclusions 114

      7.2.1   Proof: The Velocity of a Shallow Water Wave 115

  7.3   Acknowledgments 116

  7.4   References 117

8 Incorporating Active Areas of Research and Asking Complex Questions . . . 119

  8.1   The Sinking of The R.M.S. Titanic: A Case Study in Thermal Inversion
        and Atmospheric Refraction Phenomena 119

Gregory A. DiLisi

John Carroll University, University Heights, Ohio

      8.1.1   Introduction 119

	8.1.2	The R.M.S. Titanic	120
	8.1.3	Statement of the Problem	121
	8.1.4	Thermal Inversion	122
	8.1.5	Refractive Phenomena	124
	8.1.6	Classroom Demonstration	124
8.2	Conclusions		127
8.3	References		128

9   Making Local Connections **129**

9.1   Monday Night Football–Physics Decides Controversial Call: A Case Study in Observational Errors 129

Gregory A. DiLisi
John Carroll University, University Heights, Ohio

Richard A. Rarick
Cleveland State University, Cleveland, Ohio

	9.1.1	Introduction	129
	9.1.2	Statement of the Problem	131
	9.1.3	Solution	135
	9.1.4	Results and Suggestions	137
	9.1.5	And We Lived Happily Ever-After	137
9.2	General Values		138
9.3	Acknowledgments		139
9.4	References		140

10   That's a Wrap! **141**

10.1   A Case Studies Approach to Teaching Introductory Physics 142

Gregory A. DiLisi
John Carroll University, University Heights, Ohio

Richard A. Rarick
Cleveland State University, Cleveland, Ohio

Alison Chaney
John Carroll University, University Heights, Ohio

Stella McLean
John Carroll University, University Heights, Ohio

| 10.2 | Reactions to Case Studies | 142 |
| 10.3 | References | 144 |

xviii

Authors' Biographies 145

Preface

INTRODUCTORY REMARKS

My interest in case studies and forensic physics came about by accident. As I will discuss in Chapter 9, I was at a department meeting when one of my colleagues described his part-time job as a referee for local high school football games. He had me spellbound as he described his involvement in a controversy surrounding a hotly contested playoff game. A quarterback had completed a spectacular pass that would have won the game; however, my colleague ruled that the quarterback's airborne knee and foot were *beyond* the line of scrimmage when the ball was released, thus negating the play. In the parlance of referees, the play was "an illegal forward pass" [1]. After a series of official protests were lodged against him, my colleague was adamant that he had called the play correctly and that a videotape of the play would vindicate him. He made his case: "I have a videotape of the play. If only someone could scientifically analyze it for me." That kind of talk gets a physicist's blood pumping! Rest assured, my analysis of the video proved my friend had indeed made the correct call but, more importantly, it was my first taste of using forensic physics to re-examine an actual historical event (albeit a local high school football game). I was hooked!

Today, I use case studies as a way to teach not only physics, but topics in engineering, problem-solving, critical thinking, and even ethics. Indeed, the pedagogical utility of case studies is a growing and well-researched area of physics education [2–4]. For example, the University at Buffalo, with support of the National Science Foundation, hosts the National Center for Case Study Teaching in Science [5]. The Center's site contains a searchable database of over 750 peer-reviewed cases studies, encompassing all areas of science and engineering. Likewise, the American Physical Society and American Association of Physics Teachers have prepared guidelines for structuring case studies to assist teachers in achieving their desired student learning outcomes [6]. Furthermore, the American Physical Society's Forum on Education provides a nice summary of the rationale and benefits of using case studies in the physics classroom [7]. For example, in addition to re-examining historical events, case studies allow teachers to incorporate the reverse engineering of products into their classroom activities and make for powerful senior-level capstone projects in which students must build prototypes of newly conceived devices to address specific concerns of a client. The list goes on. On a final note, I should point out that lumping together the various formats in which case studies are utilized is somewhat problematic since each has its advantages/disadvantages and addresses different types of learning outcomes. However, I want to create a single label to represent the set of pedagogies that implement an up-close, in-depth investigation of a single event as a means of teaching content. Therefore, in this text, I use the phrase "case studies" in its most generic sense.

FORMAT OF THIS BOOK

Since analyzing that controversial football play, my colleagues and I have written several articles which focus on a forensics-style re-examination of other controversial historical events. In each article, we presented a case study as a pedagogy for teaching topics from both introductory- and advanced-level physics courses. Capitalizing on these articles, I recently assembled our works into a series of *"MythBusters-style"* modules for pre-service teachers enrolled in a science education course where we trial-tested the activities presented in each chapter of this book (Chapter 10 discusses the reactions of the pre-service teachers to our material). The goal of these modules was to widen the pedagogical viewpoint of the pre-service teachers by exposing them to case studies as a means of teaching physics, problem-solving, and critical-thinking. Combining our previous articles, the classroom modules for pre-service teachers, the best practices from the literature, and the lessons learned from the implementation of our prior works, we identified nine advantages of using case studies as a pedagogical approach to teaching. Each of these nine "pedagogical advantages" is the focus of a chapter of this book. Within each chapter, a case study is used as a way to highlight the "pedagogical advantage," although we emphasize that each chosen case study could be used to highlight any of the identified advantages. Each case study uses physics to analyze a controversial historical event and attempts to resolve the historical debate. Given the format just described, you might say that this book is a case study in case studies.

The advantages which are highlighted in each chapter and the accompanying case studies are summarized below.

1. **Case studies allow teachers to emphasize that scientists now take a forensics-approach to historical events.** Scientists no longer adopt a strictly passive approach to history. Instead, they bring sophisticated analytical tools to scrutinize why certain events happened. The 2015 American Football Conference Championship game between the New England Patriots and the Indianapolis Colts, better known as *"Deflategate,"* is a powerful example of how the historical record is open to re-examination and re-interpretation [8].

2. **Case studies are interdisciplinary, have broad appeal, and make personal connections to students.** Case studies were first used in the 1820s as a way of teaching the social sciences and quickly became associated with teaching anthropology, history, sociology, law, medicine, and psychology. Today, however, case studies are used in business, education, and all sub-disciplines of the STEM-fields. Our analysis of a spectacular scene from the movie *Black Panther*, demonstrates how case studies have been used to make broad, interdisciplinary appeal [9].

3. **Case studies, more so than traditional pedagogies, raise historical awareness in students and bring historical contexts to new generations of students.** By presenting stu-

dents with relevant background information, comparative timelines, and leading theories as to why events unfolded as they did, we can bring sometimes forgotten events to new generations of students. The tragic story of the *Lady Be Good,* a World War II B-24 bomber that mysteriously disappeared in 1943, is a riveting tale of courage that can be handed down to the next generation of students while simultaneously tackling multiple topics in physics and engineering [10].

4. **Case studies highlight the importance of operational definitions in scientific experiments.** With a case study, students see, perhaps for the first time, how operational definitions compare phenomena of interest against known standards or accepted protocols. Our article on the *Hindenburg* disaster provides an excellent case study to illustrate this point. On May 6, 1937, the German zeppelin *Hindenburg* caught fire while preparing to dock at the Naval Air Station in Lakehurst, New Jersey. The ensuing fire destroyed the massive airship in 35 seconds. We present the historical debate as: *"What was the source of fuel for the fire that destroyed the* Hindenburg?" [11].

5. **Case studies demonstrate the phenomenon of "Normalization of Deviance" that plagued several notorious engineering disasters.** Often, groups of scientists and engineers go to extreme lengths to test and re-test various design concepts and construction techniques. Immersed in an isolated and high-stress environment, these groups develop a false sense that their products are "fail-safe," "full-proof," and "invincible." Over time, the group accepts a lower and lower standard of performance until that standard becomes the new "norm." This phenomenon is known as "Normalization of Deviance," since the departure from a higher, more robust standard has been normalized. For this topic, a case study was designed to honor the crew of Apollo 1. On January 27, 1967, a fire swept through the interior of NASA's *"AS-204"* Command Module and killed American astronauts Roger Chaffee, Virgil "Gus" Grissom, and Edward White II during a rehearsal of their upcoming space flight. We present the historical debate as: *"What was the source of fuel for the Apollo 1 fire?"* [12].

6. **Case studies demonstrate the *"perfect storm scenario"* — how a progression of events often results in an unlikely or unforeseen outcome.** Our article on the sinking of the *S.S. Edmund Fitzgerald* provides the basis of this case study. On November 10, 1975, the Great Lakes bulk cargo freighter *S.S. Edmund Fitzgerald* suddenly and mysteriously sank during a winter storm on Lake Superior. All 29 men onboard perished. Students see that an unlikely cascade of conditions (*i.e.,* the *"perfect storm scenario"*) were met, thus placing the ship at the wrong place at the wrong time. We present the historical debate as: *"Why did the S.S. Edmund Fitzgerald sink in Lake Superior on November 10, 1975?"* [13].

7. **Case studies showcase that scientists and engineers must often develop simulations or test analog materials in lieu of actual substances — especially if those substances are**

prohibitively rare, precious, expensive, or dangerous to test. To illustrate this point, we examine the conditions of the Indian Ocean Tsunami following the Sumatra–Andaman earthquake of December 26, 2004. As a result of the earthquake, a 1,200-km × 900-km area of the ocean floor slipped 15 vertical meters where the Indo-Australian plate subducts under the smaller overriding Burma microplate. The ensuing tsunami led to the deaths of more than 200,000 people and devastated parts of Indonesia, Sri Lanka, India, Thailand, Somalia, and Burma. We present the historical debate as: *"What factors contributed to the formation of the Indian Ocean Tsunami of 2004?"* [14].

8. **Case studies allow teachers to discuss active areas of research and ask complex questions while still covering the appropriate content and maintaining the appropriate level.** Teachers are often handcuffed to teach predetermined, district-, or department-prescribed curricula. Case studies allow front-line research to be brought into the classroom. Teachers can emphasize to students the importance of staying current with recent developments in scientific research and demonstrate how this informs teaching. The sinking of the "unsinkable" R.M.S. Titanic on its maiden trans-Atlantic voyage is the hallmark example of such an opportunity. On April 14, 1912, the massive British passenger liner struck an iceberg and sank in under three hours. Over 1,500 passengers and crew perished. We present the historical debate as: *"Why did Titanic hit the iceberg in the first place?"* [15].

9. **Case studies offer teachers and students the opportunity to explore and re-examine local events.** Not all case studies have to involve historical events from the national- or international-scenes. Re-examining local events, or events that transpire at school, can bring content directly into students' lives. Thus, case studies personalize the curriculum while emphasizing how content can be used to treat a wide range of events. For this topic, we revisited a well-known controversial football play between two Cleveland-area high school football powerhouses. The play is legendary in Cleveland lore. A video analysis of the play opens up the historical debate to, *"You make the call!"* and asks students to determine if the referee indeed made the correct ruling [16].

REFERENCES

[1] The 2001 National Federation of State High School Associations Football Rulebook states: "The passer commits an illegal forward pass when he has one foot beyond the plane of the neutral zone (defined by the line of scrimmage) when he releases the ball on a forward pass. The pass is illegal. An illegal forward pass is part of a running play with the end of the run being the spot from which the pass is thrown." xix

[2] C. F. Herreid and N. A. Schiller, Case studies and the flipped classroom, *J. Coll. Sci. Teach.*, 42(5):62–66, May/June, 2013. DOI: 10.2505/4/jcst14_044_01_75. xix

[3] T. F. Slater, Teaching astronomy with case studies, *Phys. Teach.*, 53:506–507, November, 2015. DOI: 10.1119/1.4933161. xix

[4] A. Yadav, M. Lundeberg, M. DeSchryver, and K. Dirkin, Teaching science with case studies: A national survey of faculty perceptions of the benefits and challenges of using cases, *J. Coll. Sci. Teach.*, 37(1):34–38, 2007. xix

[5] The National Center for Case Study Teaching in Science, University at Buffalo, June 19, 2018. http://sciencecases.lib.buffalo.edu/cs/ xix

[6] The Joint Task Force on Undergraduate Physics Programs, Phys21: Preparing physics students for 21st century careers, *American Physical Society and the American Association of Physics Teachers*, June 27, 2018. https://www.aps.org/programs/education/undergrad/jtupp.cfm xix

[7] D. M. Katz, Using cases in introductory physics, *Forum on Education*, The American Physical Society, Summer, 2009. https://www.aps.org/units/fed/newsletters/summer2009/katz.cfm, June 5, 2018. xix

[8] G. A. DiLisi and R. A. Rarick, Bouncing back from "Deflategate," *Phys. Teach.*, 53:341–346, September 2015. DOI: 10.1119/1.4928347. xx

[9] G. A. DiLisi, Holy high-flying hero! Bringing a superhero down to earth, *Phys. Teach.*, 57:6–8, January 2019. xx

[10] G. A. DiLisi, A. Chaney, K. Kane, and R. Leskovec, The Lady Be Good: A case study in radio frequency direction finders, accepted by *Phys. Teach.*, July 2020. xxi

[11] G. A. DiLisi, The Hindenburg disaster: Combining physics and history in the laboratory, *Phys. Teach.*, 55:268–273, May 2017. DOI: 10.1119/1.4981031. xxi

[12] G. A. DiLisi and S. McLean, The Apollo 1 fire: A case study in the flammability of fabrics, with "Supplemental material for on-line appendix," *Phys. Teach.*, 57:236–239, April 2019. DOI: 10.1119/1.5095379. xxi

[13] G. A. DiLisi and R. A. Rarick, Remembering the S.S. Edmund Fitzgerald, *Phys. Teach.*, 53:521–525, December 2015. DOI: 10.1119/1.4935760. xxi

[14] G. A. DiLisi and R. A. Rarick, Modeling the 2004 Indian Ocean tsunami for introductory physics students, *Phys. Teach.*, 44:585–588, December 2006. DOI: 10.1119/1.2396776. xxii

[15] G. A. DiLisi, The R.M.S. Titanic: A case study in thermal inversion and atmospheric refraction phenomena, submitted to *Phys. Teach.*, April 2020. xxii

[16] G. A. DiLisi and R. A. Rarick, Monday night football: Physics decides controversial call, *Phys. Teach.*, 41:454–459, November 2003. DOI: 10.1119/1.1625203. xxii

Gregory A. DiLisi
July 2020

Acknowledgments

Several individuals significantly strengthened the presentation of this material. Working "behind the scenes," their efforts merit every possible recognition. Therefore, I would like to thank:

Richard Grejtak, a gifted teacher and scholar, who taught me how to write a sentence. Decades after enrolling in his courses, I still strive to emulate the excellence and excitement that he brought to each class. As a teacher, I could not ask for a better role-model.

My support team at home: **Linda DiLisi, M.D.** and **Carmela DiLisi**. Over the years, Linda and Carmela have served as volunteers (often against their better judgment) for many of the activities developed in this book. For example, Carmela was the artist for the graphic appearing in Chapter 1 as well as the inspiration and artist for the Chapter 2 analysis of the *Black Panther* movie. You can also see Carmela in Chapter 3 working the radio frequency direction finder. Linda and Carmela dropped the bouncing balls of Chapter 1 and handled the samples of fabric that you see burning in Chapters 4 and 5.

My friend, co-author, and the finest engineer walking the planet today, **Richard Rarick**. Rick's craftsmanship as an engineer and dedication to the art of teaching manifest themselves in several sections of this text. Rick's attention to detail is simply unparalleled.

My able and supportive team of publishers: **Susanne Filler** and **Melanie Carlson**, from Morgan & Claypool Publishers, who turned the idea of this book into reality; **Sara Kreisman** for the copyediting; and **Dr. C.L. Tondo** and his team at T&T TechWorks, Inc. for the typesetting.

My graduate assistants over the past several years: **Alison Chaney** and **Stella McLean**. Alison and Stella proofread many of the manuscripts assembled to produce this text.

My co-authors on Chapter 3: **Br. Kenneth Kane, C.S.C., KG8DN**, and **Robert A. Lesovec, K8DTS**. Ken and Bob possess unmatched knowledge of radio frequency direction finders. Chapter 3 simply would not be possible without their much-appreciated guidance.

Finally, I give special recognition to AIP Publishing LLC, the editorial board of *The Physics Teacher*, and the publications staff at the American Association of Physics Teachers. First, AIP Publishing LLC has graciously allowed our articles to be reprinted as part of this book. Next, at the offices of *The Physics Teacher*, **Gary White, Ph.D.** (Editor), **Karl Mamola, Ph.D.** (Professor Emeritus, Appalachian State University), and **Pam Aycock** (Managing Editor) have been wonderful colleagues who, over the years, answered countless questions and provided much-needed guidance in the development of the work presented here. Last, at AAPT, **Jane Chambers** (Senior Publications Editor) has provided editorial revisions and repairs to get our prior works into the tip-top shape you see gracing the pages of *The Physics Teacher*.

Gregory A. DiLisi
July 2020

CHAPTER 1

Taking a Forensics Approach to History

Case studies allow teachers to emphasize that scientists now take a forensics approach to historical events

Scientists no longer adopt a strictly passive approach to history. Instead, they bring sophisticated analytical tools to scrutinize why certain events happened. Far from being a set of agreed upon immutable facts, the historical record is now open to re-examination and re-interpretation. The 2015 American Football Conference Championship game between the New England Patriots and the Indianapolis Colts, better known as *"Deflategate,"* is a powerful example of how the historical record can be re-opened time after time. At halftime of this game, officials determined that 11 of the 12 footballs being used by the Patriots on offense were significantly underinflated. The team was immediately accused of intentionally using underinflated balls to give its offense an advantage. We present the historical debate as: *"Did the New England Patriots cheat during the AFC Championship game of 2015?"* Although our laboratory activity could focus on the ideal gas law and the effects of temperature on pressure, a simple analysis of the sound of bouncing balls resolves the debate. Sound waves from various professional-grade sports balls are analyzed as they bounce off of the floor. Discussions of projectile motion, conservation of energy, and linear impulse/momentum support the analysis.

1.1 BOUNCING BACK FROM "DEFLATEGATE": A CASE STUDY IN THE PHYSICS OF A BOUNCING BALL

Gregory A. DiLisi, *John Carroll University, University Heights, Ohio*
Richard A. Rarick, *Cleveland State University, Cleveland, Ohio*

1.1.1 INTRODUCTION

Halfway through the National Football League's 2015 American Football Conference (AFC) Championship game between the New England Patriots and Indianapolis Colts [1], game officials discovered that the Patriots were using under-inflated footballs on their offensive snaps (see Figure 1.1). A controversy ensued because the Patriots had actually supplied these balls

Figure 1.1: Even today, controversy surrounds the 2015 AFC Championship game between the New England Patriots and the Indianapolis Colts. The Patriots won the contest, 45 to 7, but the game quickly became known as "Deflategate." Special thanks to Carmela DiLisi for designing this graphic.

to the game's Referee just hours before kick-off. In a rare but touching display of solidarity, athletes and physicists have since agreed that using under-inflated footballs gives an unfair advantage to the offensive team since its players can improve their grip on the ball. Media outlets focused their attention on two possible culprits behind the deflationary debacle: either the Patriots had intentionally under-inflated their supply of footballs … or the climatic conditions, coupled with the various impacts to which the balls were subjected during the course of the game, had somehow altered the internal air pressure of the balls. This controversy soon became known as "Deflategate."

The purpose of this case study is to analyze the physics of "Deflategate" using the basic principles of physics covered in a typical introductory physics courses. First, we provide some background information on the actual 2015 AFC Championship game and subsequent media-blitz surrounding the controversy. This information will help us contextualize "Deflategate" as a real-word application of "physics in action." Next, we recast the spotlight on "Deflategate" from its traditional focus, the Ideal Gas Law, to a new one—the physics of a bouncing ball. We then use this scenario as a motivation for a fun but informative set of experiments that can be carried out using equipment already found in most high school or college laboratories. The subsequent data analysis relies on three basic principles: **Projectile Motion**, **Conservation of Energy**, and **Linear Impulse/Momentum**. The analysis showcases the application of introductory physics to the world of sports and demonstrates how multiple problem-solving strategies can be used to examine different aspects of a single controversy. Finally, some experimental results are presented and discussed.

1.1.2   THE CONTROVERSY

Throughout the 2014–2015 NFL football season, several teams (most noticeably the Indianapolis Colts and Baltimore Ravens) suspected that the New England Patriots were playing with under-inflated footballs while on offense. Media reports confirm that as early as November 2014, NFL officials had been alerted to the possibility that the Patriots were routinely using under-inflated footballs on offensive. With a history of alleged violations of League rules, suspicion immediately fell upon Patriots head coach, Bill Belichick, as the mastermind behind the *intentional* use of under-inflated balls. *Had Belichick purposely supplied under-inflated balls for some nefarious purpose?* Both athletes and physicists agree that the primary benefit of using an under-inflated football is that the reduced air pressure in the ball makes it easier to grip. Presumably, increasing the grip on the ball gives an advantage to the offense since the ball is now easier to throw and catch and less likely to fumble, especially in the cold and wintry conditions that often dominate the games of the NFL's post-season. Starting in 2006, the National Football League's rules require that each team supply the Referee with 12 footballs that will be used by that team while on offense. According to the 2013 Official Playing Rules of the National Football League: Rule 2 — "The Ball," Section 2 — "Ball Supply:"

> ### Rule 2 The Ball
> #### Section 2
> #### BALL SUPPLY
>
> - *Each team will make 12 primary balls available for testing by the Referee two hours and 15 minutes prior to the starting time of the game to meet League requirements. The home team will also make 12 backup balls available for testing in all stadiums. In addition, the visitors, at their discretion, may bring 12 backup balls to be tested by the Referee for games held in outdoor stadiums. For all games, eight new footballs, sealed in a special box and shipped by the manufacturer to the Referee, will be opened in the officials' locker room two hours and 15 minutes prior to the starting time of the game. These balls are to be specially marked by the Referee and used exclusively for the kicking game.*

- *In the event a home team ball does not conform to specifications, or its supply is exhausted, the Referee shall secure a proper ball from the visitors and, failing that, use the best available ball. Any such circumstances must be reported to the Commissioner.*

- *In case of rain or a wet, muddy, or slippery field, a playable ball shall be used at the request of the offensive team's center. The Game Clock shall not stop for such action (unless undue delay occurs).*

  Note: It is the responsibility of the home team to furnish playable balls at all times by attendants from either side of the playing field.

Using this process, quarterbacks get to use footballs that they have repeatedly handled and for which they have developed a "feel" that suits them. Also, the rule ensures that teams do not play with a football supplied by another team except after recovering a fumble or interception. The rules further stipulate that two hours and 15 minutes prior to every game, the Referee must measure the pressure of each of the 24 game-balls (using a pump or gauge supplied by the home club) and verify it to be between 12.5 and 13.5 PSI. Again, according to the 2013 Official Playing Rules of the National Football League: Rule 2 — "The Ball," Section 1 — "Ball Dimensions:"

Rule 2 The Ball
Section 1
BALL DIMENSIONS

- *The Ball must be a "Wilson," hand selected, bearing the signature of the Commissioner of the League, Roger Goodell.*

- *The ball shall be made up of an inflated (12 1/2 to 13 1/2 pounds) urethane bladder enclosed in a pebble-grained, leather case (natural tan color) without corrugations of any kind. It shall have the form of a prolate spheroid and the size and weight shall be: long axis, 11 to 11 1/4 inches; long circumference, 28 to 28 1/2 inches; short circumference, 21 to 21 1/4 inches; weight, 14 to 15 ounces.*

- *The Referee shall be the sole judge as to whether all balls offered for play comply with these specifications. A pump is to be furnished by the home club, and the balls shall remain under the supervision of the Referee until they are delivered to the ball attendant just prior to the start of the game.*

As a side note, the acceptable pressure of 12.5 to 13.5 PSI for each football, as mandated by NFL Rule 2 — Section 1, can be a bit misleading. Obviously, with atmospheric pressure typically accepted to be 14.7 PSI, a football would collapse if its inside pressure were truly less than one atmosphere. Instead, the League rule refers to the "gauge pressure" (or "pressure difference" between the inside and outside of the ball), not the "total pressure." Pressure gauges typically measure the pressure relative to the surrounding atmosphere. Thus, if ever using the Ideal Gas Law to tackle the physics of "Deflategate," be sure to use total pressure by adding 14.7 PSI to the gauge pressure.

Once the pressures are verified, the Referee is to store the balls until handing them over to a game-attendant just prior to kick-off. Other attributes of the footballs such as their dimensions and weights are specified by League rules, but are not required to be verified by any of the game officials. Finally, regarding the supply of game-day footballs, League rules also dictate the following: only the Referee is required to verify that the 24 game-balls comply with League specifications; the home team is to supply a set of 12 back-up balls for pressure-testing; and a separate set of 8 balls directly mailed to the Referee by the manufacturer is to be used for all aspects of the kicking game.

On Sunday, January 18, the New England Patriots routed the Indianapolis Colts, by the score of 45 to 7, in the NFL's 2015 American Football Conference championship game. The game proved to be one of the most lop-sided championship games in the history of the AFC. The game was held at Gillette Stadium, home of the Patriots, in Foxboro, Massachusetts. Air temperatures were recorded as: 52°F at kick-off (6:50 pm), 52°F at halftime, and 46.9°F at the final gun. Though the game ended in a rout, the halftime score suggested a more even contest at 17 to 7 in favor of the Patriots. Before the half ended, with the game's outcome still very much in question, a controversy erupted that continues to be the debate of athletes, sports writers, and physicists around the world. During the first half of the game, the Patriots quarterback Tom Brady, threw an interception to Colts defensive linebacker D'Qwell Jackson (see Figure 1.1).

Jackson, suspecting that the Patriots might indeed be using under-inflated balls, immediately handed the football to the Colts equipment manager for inspection. Once alerted by the Colts, game officials noticed that 11 of the 12 footballs being used in the game by the Patriots, were "significantly under-inflated." The exact measurements of pressure, as well as the temperature and time at which these measurements were taken, were not logged. News reports indicate that when checked by the Referee at halftime, the 11 balls in question were 1.4–2 PSI under the minimum pressure; however, these values of differentials in pressure have been debated. During the halftime intermission, the 11 under-inflated balls were re-inflated to proper pressure (though again, exact values of time, temperature, and pressure were not logged). Further confusion exists as to whether the set of footballs used by the Patriots to play the second half were the re-inflated balls or their set of backup footballs.

… and so began what soon became known as: "Deflategate" (the moniker makes an obvious connection to the 1970s "Watergate" scandal of political corruption). Had the Patriots *pur-*

posely under-inflated the footballs to give the team an advantage on offense … or could physics be the culprit? Perhaps the balls had simply deflated due to the changing weather conditions between the indoor location where the balls were inflated and the outdoor conditions of Gillette Stadium in mid-January. Additionally, perhaps the conditions to which the balls were submitted during the course of the first half, (i.e., throwing, fumbling, bouncing, etc.) also contributed to the change in air pressure inside the footballs? **Thus, the controversy of "Deflategate" was launched and the question on everyone's mind became: "Who was the culprit … Patriots Coach, Bill Belichick? … Patriots Quarterback, Tom Brady? … or physics?"**

1.1.3   THE MEDIA BLITZ–PHYSICS TO THE RESCUE

Within hours of the final gun, reports began to flood news outlets with speculation about the physics of "Deflategate." Op-ed pieces by columnists, bloggers, current and former athletes, and physicists are too numerous to cite; however, some of the highlights include the following.

- On Saturday, January 24, Bill Belichick held one of the most tongue-tied, pseudo-scientific press conferences since "The Professor" first lectured Gilligan on the subtleties of quantum mechanics. Attempting to deflate "Deflategate," Belichick fumbled through an explanation of how the footballs naturally became under-inflated. A particular highlight of the press conference was when Belichick mentioned that he wasn't the "Mona Lisa Vito" of air pressure (Vito refers to Marisa Tomei's Academy Award winning performance as a know-it-all auto mechanic in the movie "My Cousin Vinny."): "I would not say that I'm the Mona Lisa Vito of the football world as she was the car expertise area. All right?" In his press conference, Belichick theorized that the changing pressures in the footballs were the manifestations of two phenomena. First, "climatic conditions:"

> "We found that once the footballs were on the field over an extended period of time, in other words, they were adjusted to the climatic conditions and also the fact that the footballs reached an equilibrium without the rubbing process, that after that had run its course and the footballs had reached an equilibrium, that they were down approximately one-and-a-half pounds per square inch. When we brought the footballs back in after that process and re-tested them in a controlled environment as we have here, then those measurements rose approximately one half pound per square inch. So the net of one and a half, back to a half, is approximately one pound per square inch, to one and a half."

Next, Belichick offered the second of his two phenomena, "the rubbing process:"

"Now, we all know that air pressure is a function of the atmospheric conditions. It's a function of that. If there's activity in the football relative to the rubbing process, I think that explains why when we gave them to the officials and the officials put it at, let's say 12.5, if that's in fact what they did, that once the football reached its equilibrium state, it probably was closer to 11.5. But again, that's just our measurements."

- Saturday, January 24, "Saturday Night Live" opened its show with a sketch on the controversy.

- On Sunday, January 25, Bill Nye, appearing on "Good Morning America" took to the air and took the air out of Belichick's explanation stating that it "didn't make any sense." Nye explained: "Rubbing the football, I don't think you can change the pressure. To really change the pressure you really need one of these …" Nye then produced an inflation needle and held it in front of the camera and said: "… the inflation needle!"

- On Tuesday, January 27, League sources confirmed that the focus of the "Deflategate" investigation was on a Patriots locker room attendant who was seen on surveillance camera to take all 24 game-balls into a restroom for 90 seconds. Speculation then arose that Belichick had inflated the balls two and a half hours before game time in a heated sauna so that the Referee would indeed measure a proper air pressure at the appropriate designated time. Inflating the balls in a sauna would increase the temperature-differential, and thus the pressure-differential, to which the balls were subjected once exposed to the colder climatic conditions of Gillette Stadium.

- Throughout the time between the AFC Championship game and Super Bowl XLIX, several media outlets called for the Patriots and/or Belichick to be expelled from the Super Bowl.

- Media speculation continued to swirl as more and more reports were released confirming the propensity for professional football teams to push the letter of the law in order to gain even the slightest advantage during a game. For example, during the same time "Deflategate" was unfolding, media reports confirmed that for the last two seasons, the Atlanta Falcons had been piping in artificial crowd noise into their home field to disrupt opponents' during their offensive snaps. Also at this time, the Cleveland Browns front office was found to be repeatedly violating League rules by texting coaches during games. As lifelong Cleveland sports fans, the authors can attest to the futility of the Browns' efforts to gain a competitive edge during a game—since resuming operation in 1999, the Browns have a record, as of 2019, of 101–234. That is a winning percentage of 30%

1.1.4 A NEW FOCUS FOR "DEFLATEGATE"

In general, analyzing the physics of "Deflategate" has focused exclusively on grip—namely, that the sole advantage to using an under-inflated football is that it affords the offensive team with an improved grip on the ball. The scientific analysis has therefore focused on using the Ideal Gas Law (or the Gay-Lussac portion of the Law, $P \propto T$) to determine if the air pressure of a football inflated in a "warm," interior room can in fact decrease enough, when taken to a "cool," exterior stadium, to account for the measurements observed in the 2015 AFC Championship game (52°F in Foxboro in the middle of January can hardly be considered "frigid," as some pundits and commentators have stated). Physics laboratory activities and homework problems soon popped-up in which students use the Ideal Gas Law to either confirm or discount the alleged pressure measurements taken during "Deflategate."

Although activities involving the Ideal Gas Law make for an obvious extension of "Deflategate" into the physics classroom or laboratory, the purpose of this article is to pose an alternate scenario that allows us to focus on other aspects of mechanics. Specifically, we shift the focus of "Deflategate" from the Ideal Gas Law to the physics of a bouncing ball. At the very least, our activities can be used in conjunction with activities involving the Ideal Gas Law as a means to showcase how multiple problem-solving techniques can be used to examine different aspects of the same controversy. The intention of the authors is in no way to denigrate the reputations of the New England Patriots or their head coach Bill Belichick. Instead, our intention is simply to provide an alternative approach to discussing "Deflategate" and to pilot an accompanying set of experiments that allows us to unify basic principles of mechanics while examining an interesting, real-world application of physics to sporting events. To trial-test our activity, we presented the following problem to a small group of students enrolled in an introductory physics laboratory. After stating the problem and discussing its solution, a set of experiments was conducted by the students, under the supervision of the two authors, with the hope of incorporating this scenario into our physics laboratory sequence in later semesters.

1.1.5 STATEMENT OF THE PROBLEM

Since the controversy of "Deflategate" first erupted during the 2015 AFC Championship game, most physicists have concluded that weather and game conditions were simply not the culprit. In other words, changing climatic conditions and the typical impacts to which a football is subjected during the course of a game cannot account for a drop of 2 PSI in the internal air pressure of the ball. Therefore, we posit, for instructional purposes only, that the New England Patriots *intentionally* under-inflated their game-supply of 12 footballs to the 2015 AFC Championship game. The goal of this laboratory exercise is to find out *why might the Patriots have intentionally under-inflated their supply of game-day footballs?* In a more general sense, the purpose of this laboratory exercise is to determine *why the National Football League finds it necessary to specify the internal air pressure of the footballs used in its games.*

Until now, the leading theory has focused on grip. No one doubts that under-inflated footballs are easier to grip and therefore easier to throw and catch and less likely to fumble. So a clear advantage exists for the offensive team using under-inflated footballs. However, one might also ask: *"Is an altered grip the only advantage to using under-inflated footballs?"* What other phases of the game might be impacted by the air pressure inside the ball? An immediate thought is the kicking game—the impulse delivered by a kicker's foot to a football will certainly be affected by the ball's elasticity, (i.e., a less elastic ball will remain in contact with the kicker's foot for a longer period of time, thus increasing the energy dissipated during impact and reducing the kinetic energy delivered to the ball). However, the NFL's Rule 2, Section 2 (see above) clearly states that footballs used for the kicking game are to be directly mailed to the Referee by the manufacturer and are not supplied by either team—so the Patriots would not have under-inflated their supply of game-day footballs with the hope of altering the kicking game. The only remaining phase of the game affected by a football's internal air pressure involves its **bounce**, or **rebound**, after impact. The "rebound" of a ball refers to how it bounces after colliding with another object, like another ball, the ground, or a player's shoulder pads. Therefore, we now turn our attention to the physics of a bouncing ball. In the upcoming laboratory activity, we will determine how the internal air pressure of a ball affects its rebound. For fun, we will examine the performance of several common sports-related balls, with special attention to the professional-grade football. Our analysis relies on the application of basic principles covered in a course on basic *Classical Mechanics*.

1.1.6 THE PHYSICS OF A BOUNCING BALL

As most of us grow up playing sports, we develop an intuition of how particular types of sports-related balls rebound. We expect a super-ball to be very elastic and to rebound close to the original height from which it was dropped. Conversely, we expect a baseball to be much less elastic after impacting a hard surface such as a baseball bat, since much of the incoming energy is dissipated in the loud "crack of the bat." Many sports use a ball that must be inflated with air to some recommended pressure. Aside from the incoming velocity, this air pressure plays the most important role in determining how the ball will rebound after impact. Clearly, using an improperly inflated ball will significantly disrupt a player's ability to anticipate how that ball will recover from an impact. Examples from the world of sports are numerous: if you have ever tried to dribble an under-inflated or over-inflated basketball, you know how difficult controlling the ball can be. Since soccer players often use their heads to alter the trajectory of the ball, several studies have focused on the correlation between inflation pressure and the incidence of head injuries and concussions after impacting a player's head. In football, the rebound of the ball from the ground after a punt or fumble would be significantly altered if the football were not adequately inflated.

We intuitively understand that the combination of a ball's internal air pressure and the velocity with which it impacts a hard surface, originating from the height from which it was dropped, (i.e., the drop-height), accounts for the height of its rebound, (i.e., the rebound-height). Obviously, we expect that increasing the drop-height will increase the rebound-height. Likewise, we expect that increasing the internal pressure of the ball lowers both the energy dissipated at impact and the duration of the impact, thereby increasing the rebound-height. Therefore, the same rebound-height can be achieved with a given ball by altering various combinations of its internal air pressure and drop-height—that's why both parameters must be specified when characterizing an acceptable range of rebound-heights to which a particular ball must rebound. Although most experienced athletes develop an instinctive sense of a ball's rebound, most athletes do not know that the ball used in each sport is designed specifically to perform best in the range of heights that occur most often in that sport. For example, most NBA basketball players are 6–7 feet tall. When impacting the court (on a dribble, for example), an official NBA basketball's internal air pressure of 7.5–8.5 PSI was chosen because at that pressure, the ball dissipates very little energy and is very elastic when dribbled in this range of drop-heights. Likewise, a football is designed to perform optimally in the 12.5–13.5 PSI pressure-range. If a football is significantly under-inflated, the amount of energy it dissipates when rebounding from a surface can be greatly increased; thereby confounding the trajectory of the ball's rebound. If the Patriots practiced with under-inflated footballs, they certainly would have developed a better sense of how these footballs would rebound after impact, thus giving them an unfair advantage over their opponents.

1.1.7   PHASES OF A BOUNCING BALL AND THE COEFFICIENT OF RESTITUTION

Consider dropping a ball, from rest, from a known height. We will consider the recovery of the ball from a hard surface, but additional surfaces (like grass or artificial turf) can be added as "further explorations" to our laboratory activities. Like all good physicists, we need to make some simplifications.

- The rebound of a football off of a hard surface is complicated by its distinct shape, called the "prolate spheroid." This shape introduces a number of complicating parameters to the analysis: namely, the incoming spin, angle, and orientation of the ball relative to the surface. Unpredictable bounces are observed when elongated or non-spherical objects rebound off a hard surface [2]. This unusual behavior differs from that of a spherical ball since the normal reaction force between the ball and ground at the point of contact does not usually act along a line through the ball's center of mass. Consequently, the torque applied to an elongated ball when it rebounds depends on its orientation at impact and can be significantly larger than that on a spherical ball [3]. Our analysis is greatly simplified if we neglect the shape, spin, and orientation of the ball at impact and instead focus on a non-rotating spherical ball impacting a hard surface like the ground. Although this three-pronged approximation might seem overly simplistic at

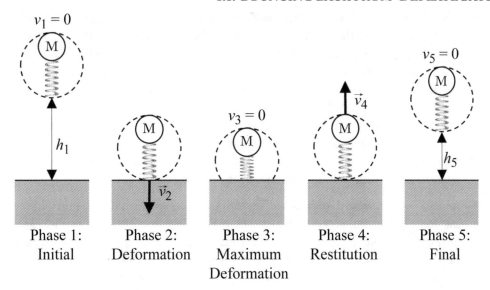

Figure 1.2: Cross-sectional view (along the long-axis of the ball) of a football rebounding from a hard surface. When viewed along its long-axis, the cross-sectional view of a football is spherical.

first, we will take steps in our experimental-design to ensure that this simplification is valid, even for the rebound of a football.

- In his head-scratching press conference, Bill Belichick referred to "the rubbing process"—his term for describing the expansion of the football over time and after repeated impacts. Many sports-balls, (i.e., soccer, football, tennis, etc.) become bigger after being used for a period of time. The football is made from a Urethane bladder enclosed in a pebble-grained, leather case. After the ball has been used, the material and stitching of the cover and linings stretch out. Additionally, stitching will loosen. Thus, the cover and linings become less capable of resisting the pressure of the air inside the bladder causing the ball to expand over time. We further simplify our analysis by making no attempt to account for the expansion of the ball after each of our experimental trials.

- Finally, we neglect air resistance.

To analyze a bouncing ball, we break up a single bounce into five distinct phases, as shown in Figure 1.2.

- *Phase 1 — Initial:* The football is released from rest from height h_1 and free-falls vertically downward under the influence of only gravity. The instant before impacting the ground, the velocity is downward as is the acceleration.

- *Phase 2 — Deformation:* The ball makes initial contact with the ground. The velocity is downward at v_2 as the ball begins to deform; however the acceleration is now upward since the ground is pushing upward on the ball with some force greater than mg. At this point, most models assume that the deformation of the ball obeys Hooke's law, (i.e., the restoring force is proportional to the displacement of the spring from equilibrium) and ignore the negligible change in gravitational potential energy that occurs over the very small distance over which the ball is deformed. Some of the kinetic energy of the ball is converted into elastic potential energy when the ball hits the ground; however, some of the kinetic energy is also dissipated during the impact due to either the internal friction of the ball, sound, vibration, or the heating of the surface.

- *Phase 3 — Maximum Deformation:* The ball continues to push the ground with a restoring force proportional to its displacement from the equilibrium position. In consequence, the ground pushes back on the ball with a force equal in magnitude but opposite in direction. The ball eventually stops as it reaches its maximum deformation. Thus, its velocity is zero, (i.e., $v_3 = 0$) while its acceleration remains upward.

- *Phase 4 — Restitution:* The ball is no longer at maximum deformation and is now pushing against the ground with a force greater than mg. The ball begins to recover and rebound into the air. The ball's velocity is upward at v_4 and the acceleration is still upward. Again, we ignore the negligible changes in gravitational potential energy in this phase.

- *Phase 5 — Final:* The ball bounces back in the upward direction. During the rebound, the stored elastic potential energy was released as kinetic energy which in turn is converted to gravitational potential energy as the ball moves up. Eventually, the ball fully rebounds and reaches its maximum height, h_5. The velocity v_5 is zero, and the acceleration is now pointing downward since the only force acting on the ball is gravity.

Often, when two objects collide (like a ball striking the ground), little information is available about the forces or processes involved in the loss of energy. For example, is the energy loss due to internal friction, the creation of sound, the generation of vibrations, or the permanent deformation of the ball or surface? Sometimes, the energy may even be stored in the ball as a result of its compression and may not be released until well after the rebound by a slow recovery of the ball to its original shape [4]. Therefore, when describing the rebound of a ball, in an effort to quantify the momentum-efficiency of the collision, physicists and sports technologists often define a coefficient e, called the "Coefficient of Restitution" (or "COR"), that measures how the linear impulse of the ball changed during its impact with the ground. Using the five phases shown in Figure 1.2, the vertical impulse experienced by the ball between phases 2 and 4 is given by:

$$\int_{t_2}^{t_4} \vec{F}(t)dt = m\left(\vec{v}_4 - \vec{v}_2\right) \Rightarrow \bar{F} \cdot \Delta t_{\text{impact}} = mv_4 + mv_2,$$

where $\bar{F}$ is the average magnitude of the impulsive force, and Δt_{impact} is the duration of the impact, (i.e., the time that the ball stays in contact with the ground). A positive sign results since the directions of $\vec{v}_4$ and $\vec{v}_2$ are opposite. The term mv_4 is called the "restitution impulse" while the term mv_2 is called the "deformation impulse." In general, the COR is defined as the ratio of the relative speeds of the two colliding objects A and B, before and after impact:

$$e = (v_B - v_A)_{after} / (v_A - v_B)_{before} .$$

Values of e have been measured for many objects striking various types of surfaces. For a perfectly elastic collision, $e = 1$ and for a completely inelastic collision, $e = 0$. Obviously, the COR depends on the elastic properties of both objects involved in a collision; however, if a relatively soft ball is dropped on a rigid surface like a hard floor, the resulting value of e provides a measure of the elastic properties of only the ball, provided there is no deformation of the surface on which it bounces [5]. Therefore, for our situation, our definition for the COR simplifies to the ratio of the restitution impulse to the deformation impulse: $e = (mv_4) / (mv_2)$. For the bounce of a generically shaped ball, including an elongated ball like a football, the COR is actually defined by the ratio of the speeds of outgoing-to-incoming contact points, (i.e., the ratio of the normal velocity components at the point of contact), not the ratio of the speeds of the outgoing-to-incoming centers-of-mass, as we have done. However, because we have simplified our problem to neglect shape, spin, and orientation, the two definitions are equivalent [3, 6, 7]. The expression for the COR can be further simplified by balancing the total energy between phases 1 and 2 as well as between phases 4 and 5. Since no non-conservative forces are involved in these phases:

$$mgh_1 = \frac{1}{2}mv_2^2 \Rightarrow v_2 = \sqrt{2gh_1}, \quad \text{and} \quad mgh_5 = \frac{1}{2}mv_4^2 \Rightarrow v_4 = \sqrt{2gh_5}.$$

Combining these results, we now have a common expression for the COR, for use with non-rotating spherically-shaped balls, impacting a hard surface: $e = \sqrt{h_5/h_1}$.

1.1.8   EXPERIMENTAL RESULTS AND DISCUSSION

To measure the height of a rebounding ball and the duration of its impact with a hard surface, one might first try simply filming a single bounce and analyzing the footage frame-by-frame. However, the speed of a typical digital camera is 30 frames per second, (i.e., 0.034 seconds per frame), while the duration of impact of a typical sports-ball with a hard surface is less than 0.01 seconds. Therefore, unless equipped with a high speed digital camera capable of film speeds greater than 100 frames per second, simple photographic analysis will not provide the temporal resolution needed for proper data analysis.

   Instead, we employ a sound sensor and data acquisition software to record the sound of several types of sports-balls bouncing one time, after striking the hard cement floor of our laboratory. This technique has been commonly used in several experiments [8–10]. First, three professional-grade sports balls were purchased: a men's basketball, football, and soccer ball. These particular balls were chosen because of the popularity of their associated sports; are easy to find at common sporting goods stores; and obviously rely on some appropriate inflation pressure (unlike solid-core types of balls like a baseball or golf ball). The costs of these professional-grade balls turned out to be the most costly expense to our experiment: the men's basketball cost $199.99, the football cost $99.99, and the soccer ball cost $159.99. Non-professional-grade versions of these balls are available at much lower costs. Next, recalling our prior list of simplifications, we need to take steps in our experimental design in order to minimize the effects of spin, shape and orientation of the ball at impact. Minimizing these effects was achieved through carefully designing an electromagnetic release-mechanism to ensure that each ball was not spinning, and was properly-oriented, when striking the ground. For the basketball and soccer ball, a small (1/2 inch × 1/2 inch) piece of flat metal was fastened to the ball—thus allowing these balls to be suspended from the electromagnet and released from rest, without rotation. Since these balls are spherical, their shape and orientation at impact were not problematic. The football proved more difficult because of its elongated shape. To minimize the effects of shape and orientation at impact, our experimental technique had to ensure that the football impacted the ground with its long axis parallel to the ground. A lightweight string was woven under the stitches of the football to create a "hanger" for the football, as shown in Figure 1.3. The hanger was then attached to a small metal binder clip. The electromagnet held the paper clip and thus suspended the football. With a little practice, this experimental technique was quite successful at releasing the football so that it impacted the ground with little rotational motion and with its long axis parallel to the ground. Overall, by using a simple electromagnet as the release-mechanism, we minimized the effects of rotation, shape, and orientation on our experimental results. For all trials, the balls were released from a predetermined height of $h_1 = 4$ feet. (Recall from Figure 1.2 that the initial drop-height is denoted by h_1.) A height of 4 feet was chosen as an approximate average value of height at which the football would most likely be carried by a running back prior to fumbling.

Next, a *PASCO* sound sensor was placed on the cement floor at the approximate location where each ball would initially impact the ground and presumably strike after its first bounce. We used the *Data Studio*TM data acquisition software (set to sample at 10 kHz) to record each impact. To start, each ball was inflated to 5 PSI below its average recommended gauge pressure. The ball was then dropped while its initial impact and subsequent bounce were recorded. This recording was repeated five times and the "best" recording kept (note that an average was not used). The ball was then inflated to 4 PSI below its average recommended gauge pressure and the initial impact and bounce were recorded five times. Again, only the best recording, as judged by the students and faculty, was kept. This process was repeated until the ball was eventually inflated to 5 PSI *over* its average recommended gauge pressure—thus, each ball was tested at 11 pressures (using increments of 1 PSI, we varied the pressure from 5 PSI below the average recommended gauge pressure to 5 PSI above the average recommended gauge pressure) with the initial impact and first bounce recorded for 5 drops at each pressure. Figure 1.4 shows all

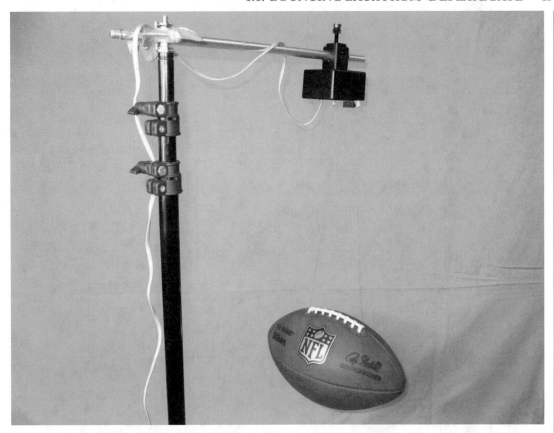

Figure 1.3: The hanger used to suspend the football is simply a lightweight string that has been woven under the stitches of the football. The string is then connected to a binder clip that is then suspended from the electromagnet.

of the equipment necessary to bring the physics of "Deflategate" to your home, classroom, or laboratory.

For each trial, data only needed to be recorded for a short time as each drop, initial impact, and bounce lasted less than two seconds. Very little practice was required to analyze the sound recordings. Since the recorded sound level makes a significant jump, relative to background noise, each time the ball impacts the ground near the sound sensor, identifying the portion of the recorded waveform corresponding to an impact is readily apparent. Typically, each recording shows: a brief interval of background noise as the ball is released from its initial height of 4 feet; a waveform corresponding to the initial impact of the ball with the ground; another brief interval of background noise as the ball arches through the air after its first bounce; and finally another waveform corresponding to the second impact of the ball with the ground. For each recording,

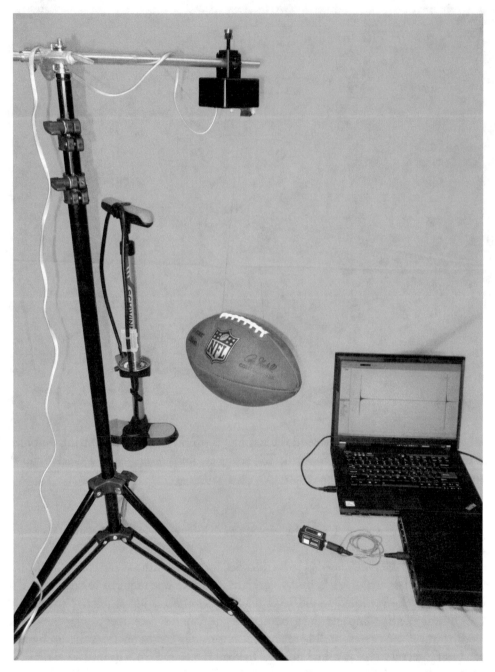

Figure 1.4: Equipment used in our experimental design. During the trials, the football was released from a height of 4 feet. The football has been lowered to fit it in the photograph.

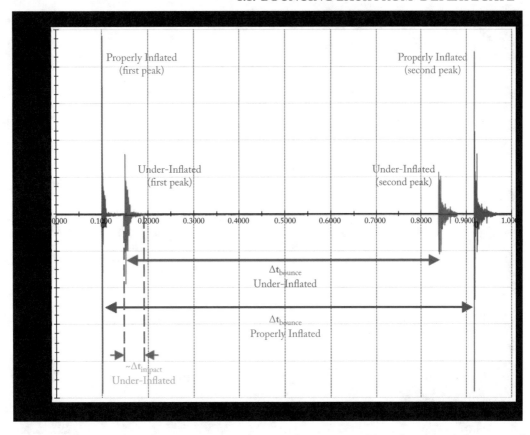

Figure 1.5: Typical plot of *Voltage* across the sound sensor vs. *Time* for a single bounce of a properly-inflated football and an under-inflated football. The determination of Δt_{bounce} for both footballs is shown.

two values were determined and tabulated: (1) Δt_{bounce}, the time between the initial impact of the ball with the ground and the time until its next subsequent impact with the ground. This value is easily determined as the time interval between initial spikes in the recorded sound levels at each impact; and (2) Δt_{impact}, the duration of the initial impact of the ball with the ground. This value is determined by estimating the width of the waveforms occurring each time the ball strikes the ground. Typical recordings are shown in Figures 1.5 and 1.6 to indicate how Δt_{bounce} and Δt_{impact} were determined for the rebound of a properly inflated football. The data for the under-inflated football (5 PSI below its average recommended gauge pressure) is shown in comparison.

From these measured quantities, the following values were then calculated: the value of the rebound-height, h_5 as: $h_5 = 1/2 * g \left(\Delta t_{\text{bounce}}/2 \right)^2$, the coefficient of restitution e as: $e =$

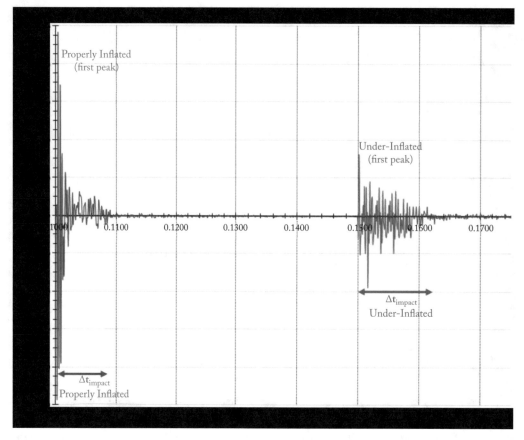

Figure 1.6: Typical plot of **Voltage** across the sound sensor vs. **Time** for the initial impact of a properly-inflated football and an under-inflated football. The determination of Δt_{impact} for both footballs is shown.

$\sqrt{h_5/h_1}$; and the percentage of energy lost due to the impact as: $U_{lost} = (h_1 - h_5)/h_1$. Table 1.1 shows our data for the professional-grade men's basketball, football, and soccer ball.

Figures 1.7, 1.8, 1.9, and 1.10 were used to discuss results with students. First, all data depicting inflation pressure was plotted using "reduced gauge pressure," $P_{reduced} = P - P_{regulation\ average}$, over the range $[-5, +5]$ PSI, (i.e., in increments of 1 PSI, the pressure of each ball was inflated from 5 PSI below to 5 PSI above its average recommended gauge pressure). For example, the recommended inflation pressure of a professional-grade football is between 12.5 and 13.5 PSI. Using the average recommended pressure of 13 PSI, the ball was first inflated to 8 PSI and tested. For each trial, the pressure was then increased by 1 PSI until a final pressure of 18 PSI was tested.

Table 1.1: Data gathered for three professional-grade sports-balls

Data Table: "Deflategate and the Physics of a Bouncing Ball"

Drop-Height, h_1 = 48"

Set	Basketball					Football					Soccer				
P_reduced	Range of Regulation Pressures: 7.5–8.5 PSI					Range of Regulation Pressures: 12.5–13.5 PSI					Range of Regulation Pressures: 8.5–15.6 PSI				
	P_regulation average: 8.0					P_regulation average: 13.0					P_regulation average: 12.0				
	Measure			Calculate		Measure			Calculate		Measure			Calculate	
	Δt_{bounce}	Δt_{impact}	h_5	COR, e	U_{lost}	Δt_{bounce}	Δt_{impact}	h_5	COR, e	U_{lost}	Δt_{bounce}	Δt_{impact}	h_5	COR, e	U_{lost}
PSI	s	s	in	PSI		s	s	in	PSI		s	s	in	PSI	
−5	0.816	0.013	32.1	0.82	33%	0.688	0.013	22.8	0.69	52%	0.670	0.014	21.65	0.67	55%
−4	0.829	0.012	33.1	0.83	31%	0.733	0.012	25.9	0.73	46%	0.720	0.013	25.00	0.72	48%
−3	0.834	0.012	33.5	0.84	30%	0.764	0.012	28.2	0.77	41%	0.750	0.013	27.13	0.75	43%
−2	0.839	0.012	33.9	0.84	29%	0.794	0.012	30.4	0.80	37%	0.780	0.013	29.34	0.78	39%
−1	0.849	0.010	34.8	0.85	28%	0.810	0.010	31.6	0.81	34%	0.790	0.012	30.10	0.79	37%
0	0.859	0.009	35.6	0.86	26%	0.818	0.009	32.3	0.82	33%	0.800	0.012	30.87	0.80	36%
1	0.869	0.009	36.4	0.87	24%	0.832	0.009	33.4	0.83	30%	0.810	0.12	31.64	0.81	34%
2	0.879	0.009	37.3	0.88	22%	0.839	0.009	33.9	0.84	29%	0.820	0.011	32.43	0.82	32%
3	0.889	0.009	38.1	0.89	21%	0.855	0.009	35.3	0.86	27%	0.830	0.011	33.22	0.83	31%
4	0.899	0.008	39.0	0.90	19%	0.866	0.008	36.2	0.87	25%	0.840	0.010	34.03	0.84	29%
5	0.909	0.008	39.9	0.91	17%	0.871	0.008	36.6	0.87	24%	0.850	0.010	34.84	0.85	27%

Under-Inflation / Over-Inflation

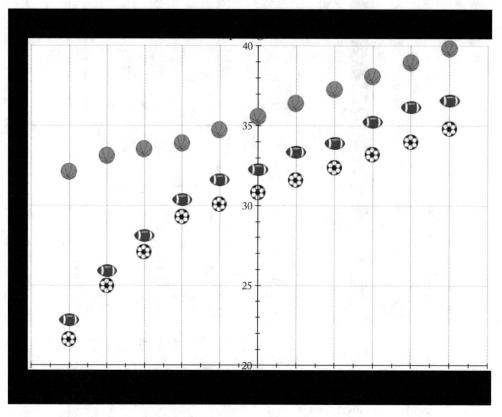

Figure 1.7: **Rebound–Height** vs. **Reduced Gauge Pressure** for three types of sports-balls.

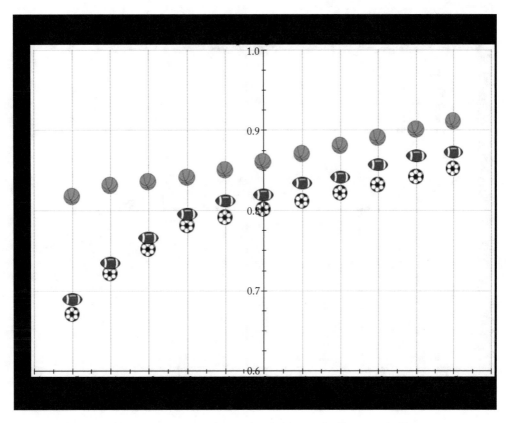

Figure 1.8: *COR* vs. *Reduced Gauge Pressure* for three types of sports-balls.

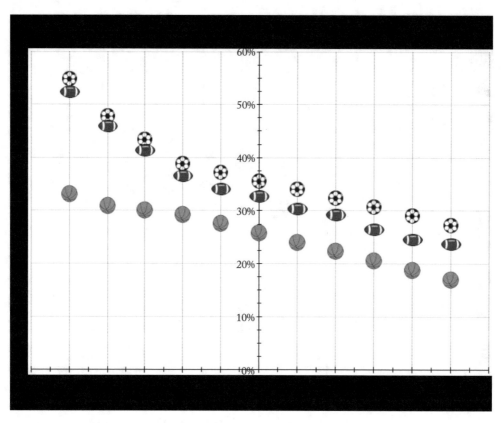

Figure 1.9: *Percentage of Lost Energy* vs. *Reduced Gauge Pressure* for three types of sports-balls.

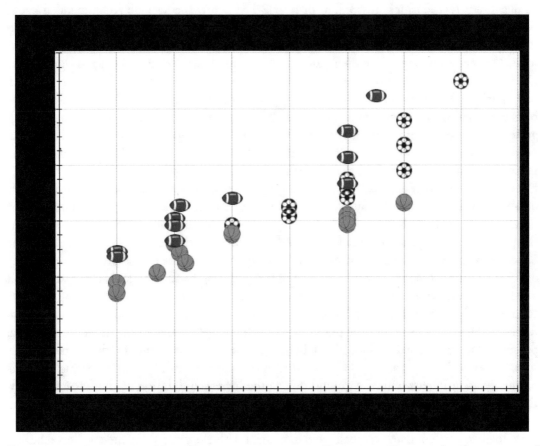

Figure 1.10: *Percentage of Lost Energy* vs. Δt_{impact}, or *Duration of Impact*, for three types of sports-balls.

Figure 1.7 shows the rebound-height of each ball vs. reduced pressure when released from a drop-height of 4 feet. As expected, the rebound-height of each ball decreases with decreasing reduced pressure. At the proper inflation pressure, the men's basketball rebounds to a height of 35.6 inches; the football rebounds to a height of 32.6 inches; and the soccer ball rebounds to a height of 30.9 inches. As a function of reduced pressure, the rebound-height of the men's basketball exhibits a linear relationship. Interestingly, the football and soccer ball do not exhibit similar linear relationships. Instead, at or near 2 PSI below the average recommended gauge pressure, the rebound-height of the football exhibits a sharp decline with respect to decreasing reduced pressure. In other words, our experimental results demonstrate that if the Patriots indeed wanted to under-inflate their game-supply of footballs in such a way as to significantly change the rebounding performance of these balls, the culprit should have under-inflated each ball by at least 2 PSI. As noted before, news reports indicated that the Patriots' game-supply of balls were indeed under-inflated in the 1.4–2 PSI range—the approximate pressure-range suggested by our experimental results at which the footballs first begin to display drastic changes in performance. After seeing this result, one student jokingly commented: "Perhaps Belichick was more versed in physics than his press conference lend us to believe?" Figure 1.8 shows the coefficient of restitution of each ball vs. reduced gauge pressure when released from a drop-height of 4 feet. Figure 1.8 supports our previous statement that each ball is designed to perform optimally in the range of heights that commonly occur in the sport for which it is intended. At the proper-inflation pressure, each ball has a coefficient of restitution in the 0.8–0.9 range: the basketball performs best with a coefficient of restitution of 0.86; the football is next with a value of 0.82; and the soccer ball last with a value of 0.80. Again, the football and soccer ball exhibit a drastic decrease in performance at or near the reduced pressure of 2 PSI, causing students to increasingly suspect intentionality behind the under-inflation of the Patriots game-supply of footballs. Figure 1.9 illustrates the percentage of lost energy, relative to the initial gravitational potential energy supplied to the ball at a height of 4 feet, as a function of reduced gauge pressure. The trend is clear: that as a ball is increasingly under-inflated, it loses increasing amounts of energy upon impact with the ground due to the increased duration of the impact. Finally, though its data is scattered, Figure 1.10 illustrates to students the expected result that increasing the duration of an impact increases the percentage of energy lost. In our trials, the over-inflated basketball and over-inflated football were in contact with the ground for the smallest duration of time, 0.008 seconds, corresponding to a loss of energy of only 17–24%, while the under-inflated soccer ball remained in contact with the ground for the longest duration of time, 0.014 seconds, corresponding to a loss of energy of 55%. Students can readily see the trend suggested by Figure 1.10 that increasing the duration of impact increases the percentage of loss of energy at impact.

1.2   CONCLUSIONS

On May 11, 2015, the NFL announced that Patriots' quarterback Tom Brady was suspended for four games, without pay, for the following season for his involvement in "Deflategate," based on "substantial and credible evidence," that he knew Patriots' employees were deflating footballs and that he failed to cooperate with the investigation. Because Brady had taken a pay cut for the 2016 season (agreeing to a $1 million salary, down from his $8 million salary in 2016), the suspension only (yes, I had difficulty using the word "only" in this sentence) cost Brady $235,000. Had Brady been suspended immediately after the controversy, his suspension would have cost him $1.9 million.

Some of life's greatest mysteries have stymied physicists for decades: "How many licks does it take to get to the center of a Tootsie Pop?" "What truly is at the center of a golf ball?" and "Why do Disney characters only have 3 fingers?" The controversy known as "Deflategate," surrounding the 2015 AFC Championship Game between the New England Patriots and Indianapolis Colts, can now be added to this list of unsolved mysteries. The purpose of this article is to add to the methods of deciphering the events of "Deflategate" using the equations and equipment of introductory physics courses. First, we provide some information on the actual events and media circus surrounding the game. Next, we provide a context for connecting "Deflategate" to the physics of a bouncing (or rebounding) ball. The analysis of a bouncing ball is ideal for the introductory physics sequence since it poses an opportunity to unite **Projectile Motion**, **Conservation of Energy**, and **Linear Impulse/Momentum** in one problem. Finally, an experimental design is constructed that allows for a simple but informative series of experiments using equipment already found in most high school or undergraduate laboratories. The key to the design is the release-mechanism that is allows students to drop various sport-related balls with no rotational motion and with the appropriate angle and orientation relative to the ground. Results can be used to illustrate a number of physical principles to students.

1.3   REFERENCES

[1] G. A. DiLisi and R. A. Rarick, Bouncing back from "Deflategate," *Phys. Teach.*, 53:341–346, September 2015. DOI: 10.1119/1.4928347. 1

[2] R. Cross, The fall and bounce of pencils and other elongated shapes, *Am. J. Phys.*, 74:26–30, 2006. DOI: 10.1119/1.2121752. 10

[3] R. Cross, Bounce of an oval shaped football, *Sports Tech.*, 3:168–180, 2010. DOI: 10.1080/19346182.2011.564283. 10, 13

[4] R. Cross, The bounce of a ball, *Am. J. Phys.*, 67:222–227, 1999. DOI: 10.1119/1.19229. 12

[5] R. Cross, The coefficient of restitution for collisions of happy balls, unhappy balls, and tennis balls, *Am. J. Phys.*, 68:1025–1031, 2000. DOI: 10.1119/1.1285945. 13

[6] H. Brody, R. Cross, and C. Lindsey, *The Physics and Technology of Tennis*, Racquet Tech Publishing, Solana Beach, CA, 2002. 13

[7] A. Nathan, Characterising the performance of baseball bats, *Am. J. Phys.*, 71:134–143, 2003. DOI: 10.1119/1.1522699. 13

[8] J. Njock-Libii, Using *Microsoft Windows* to compare the energy dissipated by old and new tennis balls, *Proc. of the National Conference and Exposition of the American Society for Engineering Education*, paper AC 2010 page 269, Louisville, KY, 2010. 13

[9] S. K. Foong, D. Kiang, P. Lee, R. H. March, and B. E. Paton, How long does it take a bouncing ball to bounce an infinite number of time?, *Phys. Ed.*, pages 40–43, January 2004. DOI: 10.1088/0031-9120/39/1/f16. 13

[10] C. E. Aguiar and F. Laudares, Listening to the coefficient of restitution and the gravitational acceleration of a bouncing ball, *Am. J. Phys.*, 71:499–501, 2003. DOI: 10.1119/1.1524166. 13

CHAPTER 2

Having Interdisciplinary Appeal

Case Studies are interdisciplinary, have broad appeal, and make personal connections to students

Case studies were first used in the 1820s as a way of teaching the social sciences and quickly became associated with teaching anthropology, history, sociology, law, medicine, and psychology. Today, however, case studies are used in business, education, and all sub-disciplines of the STEM-fields. For instance, case studies can be used to teach foundational principles and problem-solving strategies of physics, conceptual physics, multiple courses in STEM, and even engineering ethics. Case studies can serve as technologically challenging capstone projects, modules in advanced laboratory courses, and the basis of independent studies. Additionally, students may one day encounter case studies if they pursue careers in some of the disciplines mentioned above—medicine, law, or business—fields in which case studies are more commonly used than in the STEM fields. The take-home message here is that case studies have an interdisciplinary appeal with a unique ability to make direct connections to students' lives. In this chapter, we re-examine an event that showcases the application of physics to the world of pop culture (see Figure 2.1). We provide a good example of how a case study may appeal to students who are more disposed to a liberal arts focus than to STEM.

2.1 HOLY HIGH-FLYING HERO! BRINGING A SUPERHERO DOWN TO EARTH: A CASE STUDY IN UNIFORMLY ACCELERATED MOTION

Gregory A. DiLisi, *John Carroll University, University Heights, Ohio*

2.1.1 INTRODUCTION

I am always looking for ways to bring current events into my introductory physics classroom or laboratory [1]. I am especially interested in finding examples where basic principles of physics can be used to cast skepticism on assertions made by celebrities, politicians, or professional athletes. The other day, one such example literally fell into my lap.

Figure 2.1: In this chapter, our analysis brings movies and television commercials directly into the classroom.

While watching television with my 13-year-old daughter, a commercial aired that showcased both the 2018 Lexus LS 500 sedan and Marvel Studios' blockbuster superhero movie, "Black Panther." The Lexus/Marvel spot starts with the Black Panther employing a variety of astonishing acrobatics to take down a handful of thugs. At the conclusion of the fight, the superhero realizes that his alter ego is late for an important social engagement. Using a communication device built into his mask, he asks his sister to dispatch his car to pick him up, saying: "Is my ride ready?" The sister responds: "Yes … but you have to hurry!" Apparently, the superhero is so late for his social engagement that the car cannot even stop to pick him up. At this point, a remote-controlled 2018 Lexus LS 500 sedan starts screeching its way through city streets to pick up the superhero. The car speeds down a street at 54 mph (the commercial actually shows the car's speedometer at 54 mph) where the Black Panther awaits atop an overhanging bridge. At just the right instant, the Black Panther stands, drops from the bridge, free-falls through the car's fully-opened sun roof, and comes to rest comfortably in the driver's seat of the car. The commercial can be seen by simply Googling "Lexus LS 500 Commercial" or visiting YouTube [2].

Figure 2.2: Introducing "The Hooded Llama." Special thanks to Carmela DiLisi for designing this graphic.

My daughter has seen her fair share of superhero movies so I expect her to be somewhat immune to the unbelievable, physics-defying, CGI-generated special effects often saturating these movies. However, this commercial grabbed my attention because the instant it aired, she protested out loud: "That's totally impossible!" Her reaction caused me to wonder if my college students would likewise find the commercial "impossible" to believe. Here was an excellent opportunity to incorporate current events into my classroom. Could someone use basic principles of physics, covered in an introductory Classical Mechanics course, to cast doubt on the events portrayed in the commercial?

2.1.2 IT'S A BIRD… IT'S A PLANE… IT'S THE HOODED LLAMA!

To avoid any references to actual car brands and superheroes, I cast the problem using a fictitious superhero, "The Hooded Llama" (see Figure 2.2) and his equally fictitious sports car, "The Nexus 500."

Figure 2.3: The Hooded Llama awaits his ride from an overhanging bridge. Special thanks to Carmela DiLisi for designing this graphic.

My daughter has drawn the Hooded Llama performing the exact same stunt, from the same points-of-view, as depicted in the commercial: we see the Hooded Llama awaiting atop an overhanging bridge (see Figure 2.3), dropping from the bridge (see Figure 2.4), free-falling toward the car (see Figure 2.5), and falling through the car's fully opened sun roof (see Figure 2.6). In this problem, the Hooded Llama is simply trying to jump through the opened sun roof of his Nexus 500 sports car.

2.1.3   STATEMENT OF THE PROBLEM

Our superhero, of height $H_{Hooded-Llama}$ and mass $M_{Hooded-Llama}$, stands atop a bridge of height H_{Bridge}. At the correct moment, he drops from a bridge so that he falls directly into his speeding Nexus 500, traveling at constant velocity, V_{Car}. The sun roof has an opening along the direction of motion of D_{Roof}. Note that the dimension of the sun roof, along the long-axis of the actual Lexus LS 500, depends on the style (standard or panoramic) but a typical value is 18.7 inches. Also, the headroom of the car, (i.e., the distance from the floor of the car to the interior ceiling) is listed as 37.3 inches [3]. Assume the Hooded Llama drops from rest and remains upright as he passes through the sun roof, using his legs to halt his fall (see Figure 2.5). In reality, our hero would have to bend his waist to properly land in the driver's seat. Thus, our hero's buttocks, and not his legs, would stop his fall. However, in the interest of simplifying our calculations, and

Figure 2.4: The Hooded Llama prepares to jump. Special thanks to Carmela DiLisi for designing this graphic.

Figure 2.5: The Hooded Llama free-falls. Special thanks to Carmela DiLisi for designing this graphic.

Figure 2.6: The Hooded Llama passes through the sun roof of his remotely controlled car. Special thanks to Carmela DiLisi for designing this graphic.

to save our hero the indignity of further discussing his buttocks, we assume the car has enough room so that the Hooded Llama stops his fall by using his legs against the floor of the car, after which he smoothly settles into the driver's seat (see Figure 2.7).

2.1.4 MEANWHILE, BACK IN THE PHYSICS LABORATORY...

The following three questions were posed to students: (a) With what minimum speed must our hero reach the sun roof in order to pass safely through its opening? (b) From what height must he have started his fall? (c) What force would his legs have to absorb in order to come to rest in the driver's seat? Assume a mass of 200 lb for our hero. The problem combines simple applications of kinematics, uniformly acceleration motion, and the work-energy principle.

To answer (a), we start by calculating the time needed for the Hooded Llama to pass through the sun roof. Obviously, the superhero accelerates as he free-falls through the sun roof; thus, we should take into account his acceleration over the length of the fall, (i.e., the height of the superhero). However, this acceleration makes such a negligible contribution to the Hooded Llama's final velocity (it is well below a 1% contribution), we can assume a constant velocity as he passes through the sun roof. Thus, his final velocity is computed as: $t_{through-sun-roof} = (H_{Hooded-Llama}/V_{Hooded-Llama})$, where $V_{Hooded-Llama}$ is the downward velocity of the hero when his feet reach the front-edge of the sun roof. However, the sun roof will only

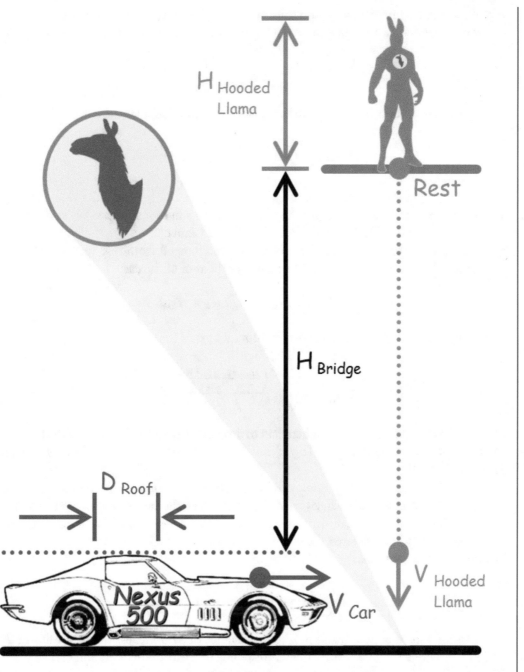

Figure 2.7: Statement of the problem. The Hooded Llama is late for a social engagement and jumps into his Nexus 500 sports car while it is moving at constant velocity.

pass underneath him for a maximum time of $t_{Max} = (D_{Roof}/V_{Car})$. Therefore, the condition under which Hooded Llama can pass safely through the sun roof is, $t_{through-sun-roof} < t_{Max}$, or:

$$\frac{H_{Hooded-Llama}}{V_{Hooded-Llama}} < \frac{D_{Roof}}{V_{Car}}.$$

This expression gives us the hero's minimum speed when he reaches the sun roof as:

$$V_{Hooded-Llama} > \frac{H_{Hooded-Llama} * V_{Car}}{D_{Roof}}.$$

Plugging in the values: $H_{Hooded-Llama} = 6$ ft, $V_{Car} = 54$ mph (or 79.2 ft/sec), and $D_{Roof} = 18.7$ in, we see our hero has to enter the opening of the sun roof at a speed of at least 208 mph (or 305 ft/sec). **"Holy High-speed Hazard, Hooded Llama!"**

To answer (b), we assume the Hooded Llama jumped from rest from the bridge, thus making his downward velocity when he arrives at the roof of the car:

$$V_{Hooded-Llama} = \sqrt{2 * g * H_{Bridge}}.$$

Combining this result with our safety condition, we get:

$$H_{Bridge} > \frac{\left(\frac{H_{Hooded-Llama}*V_{Car}}{D_{Roof}}\right)^2}{2g},$$

and that the height from which our hero had to drop as approximately 1,446 ft. The height of the Willis Tower (formerly the Sears Tower) in Chicago is 1,450 ft. **"Holy High-rise Headache, Hooded Llama!"**

Finally, we answer (c) by employing the work-energy principle. As our hero came to rest in the driver's seat, his legs had, *at most*, the distance of the headroom of the car over which to exert a force to stop his fall. In this small distance, his legs must have done work equal to his kinetic energy as he entered the sun roof. We use an average force over the distance of the headroom of the car to derive:

$$F_{avg} * d_{Headroom} = \left(\frac{1}{2}mV^2\right)_{Hooded-Llama}.$$

Using a mass of 200 lb and headroom distance of 37.3 inches, we get a value of F_{avg} of 413,915 Newtons (or 93,051 lb)—that corresponds to a vertical deceleration of 2,070 m/s^2 (or 211 g's)! Typical aircraft pilots, using modern g-suits, can only handle 9 g's; but we are talking about the Hooded Llama, not some ordinary human. **"Holy High-flying Hospitalization, Hooded Llama!"**

2.2 ACKNOWLEDGMENTS

The author gratefully acknowledges his daughter Carmela, for knowing a physics-defying commercial when she sees one. The author also acknowledges the reviewers of this manuscript and the editors of *The Physics Teacher* who significantly strengthened the presentation of this work.

2.3 REFERENCES

[1] G. A. DiLisi, Holy high-flying hero! Bringing a superhero down to earth, *Phys. Teach.*, 57:6–8, January 2019. 27

[2] Lexus LS 500 F Sport | Marvel Studios' Black Panther TV Commercial, February 21, 2018. https://www.youtube.com/watch?v=jQhsXd9qnA8 28

[3] The 2018 Lexus© LS 500, February 21, 2018. http://www.lexus.com/models/LS/specifications 30

CHAPTER 3

Raising Historical Awareness and Bringing History to New Generations

Case studies, more so than traditional pedagogies, raise historical awareness in students and bring historical contexts to new generations of students

By presenting students with relevant background information, comparative timelines, and leading theories as to why events unfolded as they did, we can bring sometimes forgotten events to new generations of students. As teachers, we are always looking for ways to bring historical events and real-world applications of physics to our students while still covering appropriate topics. Re-examining a major historical event is a powerful, sometimes forgotten, means of piquing students' interest, particularly those who are more disposed to a liberal arts focus. In this chapter,

Figure 3.1: The Consolidated B-24D *"Lady Be Good,"* [1].

we investigate the disappearance of the *Lady Be Good*, a B-24 World War II bomber that mysteriously vanished on April 3, 1943 (see Figure 3.1). The anniversary of plane's disappearance can be used to bring it to the attention of new generations of students; namely, those enrolled in our courses. This anniversary is a somber, but powerful way of raising historical awareness in our students while still adhering to the typical physics curriculum. For example, teachers can provide students with background information on the plane's crew, final voyage, and disappearance. Pictures and eye witness testimonies can be used to more artistically portray the accompanying historical events.

3.1   THE *LADY BE GOOD*: A CASE STUDY IN RADIO FREQUENCY DIRECTION FINDERS

Gregory A. DiLisi, *John Carroll University, University Heights, Ohio*
Alison Chaney, *John Carroll University, University Heights, Ohio*
Br. Kenneth Kane, C.S.C., KG8DN, *Gilmour Academy, Gates Mills, Ohio*
Robert L. Leskovec, K8DTS, *RALTEC®div GENVAC Aerospace, Highlands Hts., Ohio*

3.1.1   INTRODUCTION

Over the past several years [2], we have contributed articles to *TPT* that focus on forensics-style re-examinations of significant historical events [3–8]. The purpose of these articles is to afford students the opportunity to apply basic principles of physics to unsolved mysteries and potentially settle the historical debate. Our articles on the sinking of the *S. S. Edmund Fitzgerald* [6], the *Hindenburg* disaster [7], and the *Apollo 1* fire [8] provide especially powerful illustrations of this approach. In our most recent article, we assembled the lessons learned and best practices of our activities into a formalized pedagogy for teaching topics in physics, engineering, problem-solving, critical thinking, and ethics [9]. Adding to our repertoire of case studies, we now present the puzzling and tragic story of the *Lady Be Good*, a World War II B-24 bomber that mysteriously disappeared in 1943.

The case presents several teaching opportunities for instructors of introductory-level courses in physics and engineering. The goal of this article is to showcase classroom and laboratory exercises which we piloted with a cohort of undergraduate students enrolled in an interdisciplinary science course. Using our case study approach, students built radio frequency direction finders and used them to navigate from various locations on campus to a *"home base."* Our goal was to provide students with an easy-to-build circuit and to see if navigating home would be as easy as students might initially think.

Figure 3.2: The crew of the *Lady Be Good*. Left-to Right: Hatton, Toner, Hays, Woravka, Ripslinger, LaMotte, Shelley, Moore, and Adams.

3.1.2 THE MYSTERIOUS DISAPPEARANCE OF THE *LADY BE GOOD*

On April 4, 1943, the B-24 Liberator, *Lady Be Good* (LBG), took off from Soluch Air Base, located south of Benghazi, Libya. The new plane, named by the crew who had flown it from the U.S. to Soluch, had just joined the 376th bomber group on March 25. The nine-man crew was under the command of Pilot, First Lieutenant William J. Hatton and was making its first combat mission together. The crew (see Figure 3.2) consisted of:

- First Lieutenant William J. Hatton, Pilot

- Second Lieutenant Robert F. Toner, Co-Pilot

- Second Lieutenant Dp Hays, Navigator

- Second Lieutenant John S. Woravka, Bombardier

- Technical Sergeant Harold S. Ripslinger, Flight Engineer/Gunner

Figure 3.3: The planned mission of the LBG. After bombing the harbors of Naples, the crew was to fly to an area northwest of Benina Tower. Here, a RF beacon would guide the crew home. Map courtesy of Google Maps.

- Staff Sergeant Robert E. LaMotte, Radio Operator/Gunner

- Staff Sergeant Samuel E. Adams, Tail Gunner

- Staff Sergeant Guy E. Shelley Jnr, Waist Gunner/Assistant Engineer

- Staff Sergeant Vernon L. Moore, Waist Gunner/Assistant Radio Operator

The mission was labeled *"Mission 109"* and called for the LBG, along with 24 other planes, to bomb the harbors of Naples, Italy in a two-wave high-altitude attack. Twelve bombers would form the first wave followed by a second wave of the remaining 13. The LBG was to be one of the last in formation of the second wave. After dropping their bombs, the crews were expected to fly their B-24s southeast for a pre-set amount of time. Based on anticipated wind conditions and agreed-upon flight speeds and altitudes, the planes would arrive northwest of the air base. Here, navigators would locate a radio frequency bearing, originating from Benina Tower 30 miles north of Soluch, to guide the planes home (see Figure 3.3).

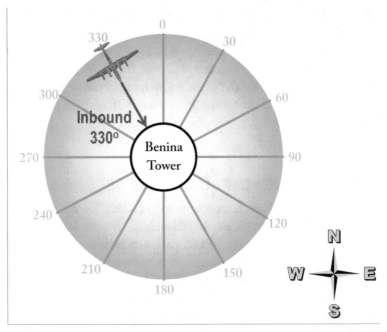

Figure 3.4: Request for help. At 12:12 am, Hatton requested an inbound emergency bearing. Benina Tower responded with a heading of 330°.

Problems plagued the mission from the very beginning, most of them involving weather conditions and strong headwinds. The LBG was one of the last B-24s to depart, getting airborne at 2:15 pm. After taking off, high winds and a sandstorm obscured visibility. Debris from the sandstorm caused nine of the bombers in the second wave to quickly abort the mission and return to base. The LBG was far behind the remaining three bombers so was unable to join them in formation. Stronger than anticipated headwinds caused the LBG to arrive at its target at 7:50 pm, much later than planned. Unable to see the targets at Naples, the crew dumped its bombs in the Mediterranean (to save fuel) and headed back to Soluch. Hatton turned the plane along the pre-planned southeasterly heading and, with the plane's Automatic Direction Finder (ADF) malfunctioning, relied on his rookie navigator to hone-in on the RF beacon that would guide the crew home. The LBG was now returning to base alone and in the dark. At 12:12 am, Hatton was lost. He radioed Benina that his ADF was not working and asked for an inbound emergency bearing. The station reported: *"Bearing is three-three-zero,"* [10]. Hatton acknowledged the message and signed out (see Figure 3.4).

This would be the last communication received from Hatton—the massive bomber vanished without a trace! The next morning, rescuers conducted an all-out search. Investigators theorized that a German night fighter had picked up Hatton's call for a bearing, honed-in on the LBG, and shot it down somewhere in the Mediterranean Sea. However, no life rafts nor any other evidence of a water crash, such as oil slicks or floating debris, were found. The search was abandoned and next of kin notified. On April 5, 1944 (exactly one year later), a board of officers declared the crew, *"missing in action and presumed dead."* What happened to the *Lady Be Good*? How could this plane, carrying the latest equipment in communication/navigation and manned by a highly trained crew of nine, simply disappear? The fate of the *Lady Be Good* would remain a mystery for almost 16 years.

3.1.3   RADIO FREQUENCY DIRECTION FINDERS

Today, localizing AM and FM sources is a popular sport (called *"fox hunting"*) for amateur radio operators and clubs. During a fox hunt, participants search for the location of a *"fox,"* a hand-held RF transmitter or other transmission source, such as a weather station. The sport also provides powerful teaching opportunities for courses in introductory physics. In the classroom, the theory of operation behind radio frequency directional finders (RDFs) is a great way to teach a wide variety of concepts: signal phase, circuit-design, spherical trigonometry, latitude/longitude, bearing, the great-circle distance between two points, equirectangular projection, and the distinction between *"parallel"* and *"antiparallel."* In the laboratory, building an RDF is a fantastic exercise that allows students to gain hands-on experience with antennas, receivers, and circuit-construction. Such a laboratory activity is ideal for an introductory electrical engineering module or soldering exercise in an introductory circuits class.

Early direction finders used a rotatable low frequency *"loop antenna."* To determine bearing, an operator would turn the loop until a null in the transmission signal was detected, indicating that the loop's axis was aligned with the direction of the signal. Since then, many different techniques have been developed. One popular type of RDF works on the principle of *"time-difference-of-arrival"* whereby two antennas are alternately connected to the input of a radio receiver. Switching between the two antennas, *at an audio frequency*, allows the user to *hear* any phase errors in the incoming signal. For example, switching between the antennas at 500 times per second, the phase difference will be detected by an FM receiver as a 500 Hz tone (approximately a B_4 note in scientific pitch notation). By rotating the RDF's antennas, one of two scenarios will play out.

A. If one antenna is slightly closer to the transmission source than the other, it receives the wave front slightly earlier in time than the other. Thus, the RF signal will have a different phase at each antenna. Since the RDF is switching between the two antennas, the switching action imposes phase modulation on the incoming signal. This phase modulation is detected by the receiver and is heard as an audio tone equal to the switching frequency (see Figure 3.5).

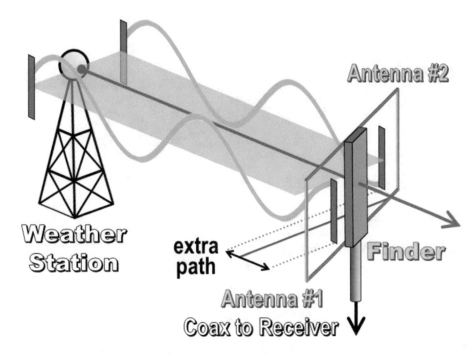

Figure 3.5: A. The theory of operation of a RDF. One antenna is closer to the transmission source than the other (indicated by the extra distance the one wave must travel); therefore, the signal has a different phase at each antenna. An audio tone is heard if the finder switches between the two antennas at an audio frequency.

B. If both antennas are the same distance from the transmission source, (i.e., if the plane of the two antennas is perpendicular, or *"broadside,"* to the direction of the signal), the antennas detect the same RF phase, so the audio tone disappears (see Figure 3.6).

Thus, students can use a RDF to locate the direction of a signal by first tuning the radio receiver to the frequency of the source, then rotating the RDF's antennas until the audio tone disappears—at that instant, the antennas must be perpendicular to the transmitter.

This type of direction finder offers several advantages for use in an introductory physics course [11, 12]. First, to determine the bearing to a transmission source, students simply rotate the finder and listen for a null in tone, rather than a peak. This null is sharp and much easier to detect than the peak from a directional gain antenna, such as a Yagi. Second, when students null the superimposed audio, they are not nulling the carrier signal. The problem with carrier-null (the principle behind a conventional loop antenna) is that as the student tunes closer to the null, the signal he/she is trying to null is getting harder to hear! With this type of direction finder, students can still hear the audio coming from the source as they null the superimposed audio.

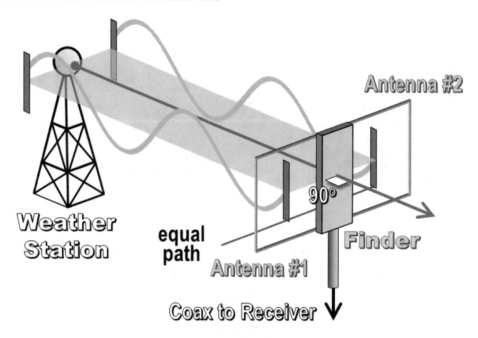

Figure 3.6: B. The theory of operation of a RDF. Both antennas are the same distance from the transmission source; therefore, the signal has the same phase at each antenna. No tone is heard, indicating that the source is perpendicular to the plane of the two antennas.

Third, since audio is being nulled, students need not watch a dial. Nulling can be done while visually surveying landscape, walking, or taking notes. Last, since this method uses phase information, it works well with strong signals so no attenuator is required. Despite their advantages, these finders offer a significant disadvantage in pinpointing the direction of an RF source—they exhibit what is known as a *"180°-ambiguity."* In general, bearing lines are inherently ambiguous because an infinite line can indicate two possible directions, 180° apart. Since students hear a null tone when both antennas are oriented 90° with respect to the transmission source, the tone will disappear when the two antennas are parallel or antiparallel to the line of transmission. Thus, students will hear two *"null points"* which are exactly 180° apart. Students cannot determine if they are pointing toward, or away from, a transmitter! The addition of a *"sense"* antenna would eliminate the 180°-ambiguity, but is not needed for our activity.

3.1.4   STATEMENT OF THE PROBLEM

We piloted our activity on a cohort of undergraduate students enrolled in an interdisciplinary science course. We divided our activity into three phases labeled: *"The Mystery of the Lady Be Good,"* *"Constructing Your RDF,"* and *"Navigating Home."* In the first phase, we presented students with the background information on the disappearance of the *Lady Be Good* that was described at the beginning of this article. We used excerpts from eyewitness accounts and archival footage to authentically portray the event. We used this case study approach in order to raise historical awareness in our students, demonstrate the *"perfect storm scenario,"* (i.e., how a progression of events often results in an unforeseen outcome), and showcase that we would be using forensic physics to solve the mystery surrounding an actual historical event. We then discussed the theory of operation of a RDF, **but did not discuss the** *"180°-ambiguity"* **inherent to its design!** In the second phase, teams of students built RDFs over the timespan of two 2-hour laboratory periods (units can also be purchased partially or fully assembled). Most teams required additional time outside of class to complete their RDFs. A number of handheld RDFs are available for purchase as kits. Two popular models are: *"The Searcher"* (available from Rainbowkits for $44.95) [13] and the *"HANDI-Finder®"* (available from RALTEC® Electronics for $33.95) [14]. We purchased RALTEC® Electronics' HANDI-Finder® since the manufacturer was local and because of the company's excellent reputation for customer service. The HANDI-Finder® works from 400–1500 Hz and has two open-loop antennas made of coat-hangar wire. The electronic switching of a CD4047B CMOS integrated chip alternately connects each antenna to a coax cable down-lead going to the antenna input of an FM receiver (purchased separately) tuned to the frequency of interest. The unit uses low power (the total current draw is only 1.7 mA from an onboard 9 V battery), is easy to build, and is inexpensive (see Figure 3.7).

   In the final phase, students downloaded a compass-app to their cell phones. Any app that determines location as a longitude and latitude (φ, λ), as well as bearing (β), will suffice. Many such apps are available for free, like *"Compass"* (version 4.9.6) from PixelProse SARL or *"Compass"* (version 2.1.2) from gabenative. Students then walked to several remote locations, chosen by the instructors to cover the full extent of our campus. Once at their locations, students used their RDFs to pinpoint the direction of a radio signal broadcast by a nearby weather station of the National Oceanic and Atmospheric Administration. The Chesterland, Ohio antenna (Call letters: KHB-59) is 12 miles from campus and broadcasts at 162.550 MHz. Students were not told the location, nor name, of the weather station so as not to bias their search for the signal. Students were only given the station's broadcasting frequency and told that finding the direction of its signal was a *"life or death"* determination since the bearing would be used to guide them to *"home base."* While outdoors, students had no difficulty in hearing the nulling of the audio tone. Students experienced difficulty in determining accurate bearings if tall buildings, large objects, and towers were nearby. Also, these RDFs did not reliably work indoors. Our recommendation is for students to take several measurements at different locations, a few feet apart, before settling on a final location and bearing. Using their compass-apps, students then recorded their locations ($\varphi_{student}$, $\lambda_{student}$), as well as their determination of the bearing to the station ($\beta_{student}$). Once all the students returned to the classroom, we collectively compared the bearings measured by students vs. the true bearings. Knowing the coordinates of the weather station ($\varphi_{station}$, $\lambda_{station}$), and those of each student, a *"true"* bearing (β_{true}), can be calculated. For small distances,

Figure 3.7: **A** RDF in action. A student uses the *"HANDI-Finder®"* to pinpoint the direction of a weather station. These finders work best is wide open, flat spaces. Nearby tall buildings hinder the finder's accuracy.

Pythagoras' theorem can be used on an equirectangular projection; however, the calculations for a spherical earth are straightforward [15–18]:

$$\beta_{true} = \text{atan2}\,(x, y)\,, \quad \text{where:}$$

$$x = \cos\,(\varphi_{station}) \cdot \sin\,(\Delta\lambda)\,,$$

$$y = [\cos\,(\varphi_{student}) \cdot \sin\,(\varphi_{station})] - [\sin\,(\varphi_{student}) \cdot \cos\,(\varphi_{station}) \cdot \cos\,(\Delta\lambda)]\,, \quad \text{and}$$

$$\Delta\lambda = \lambda_{station} - \lambda_{student}.$$

A number of sites provide code allowing the user to quickly input coordinates, determine distances and bearings, and display results directly onto a satellite map [19].

Table 3.1 shows a sampling of 5 (of 40) data points from our activity. Cells highlighted in yellow indicate that the student determined the bearing to be directed away from the weather station while cells highlighted in red indicate the student determined the bearing to be directed toward the station. Figure 3.8 shows a representation of our results on a regional map while

Table 3.1: Radio frequency direction finder

Station: National Oceanic and Atmospheric Administration—Chesterland, Ohio					
$\varphi_{station}$			$\lambda_{station}$		
Latitude			Longitude		
D:M:S			D:M:S		
41:31:21.91			-81:19:42.69		
NO	Measurements by Students			True	Difference
	$\varphi_{student}$	$\lambda_{student}$	$\beta_{student}$	β_{true}	$\beta_{student}-\beta_{true}$
	Latitude	Longitude	Bearing	Bearing	
	D:M:S	D:M:S	Degrees	Degrees	Degrees
1	41:29:22.87	-81:31:52.49	260	77.6	182
2	41:29:21.77	-81:31:55.72	75	77.6	-3
3	41:29:23.62	-81:31:54.31	270	77.7	192
4	41:29:26.64	-81:31:50.52	90	78.0	12
5	41:29:25.88	-81:31:47.67	66	78.9	-12

Figure 3.9 provides a more detailed representation on a campus map. Figure 3.8 also depicts how the instructors chose to send students to locations that covered the full extent of our campus. Encouragingly, results demonstrate that all students were capable of locating the direction of the RF signal to within 15°, either side of the true bearing. The take-home message here is that the RDFs work … and work well! As expected, many students (25%) fell victim to the *"180°-ambiguity"* inherent to the design of these finders. Initially, we expected ∼50% of the students to fall victim to the ambiguity since students presumably have a 50% chance of orienting themselves antiparallel to the line of transmission. This discrepancy can be understood by the layout of our campus. As students exit the science center, they naturally face toward the direction of the weather station. Instructors noticed that this layout biased the students to search forward for the transmission signal. Interestingly, the more curious students, who performed a more thorough 360°-sweep of the campus, were the ones who typically fell victim to the *"180°-ambiguity."* Finally, no student found two nulls in tone.

3.1.5   MYSTERY SOLVED: *"THE* 180°*-AMBIGUITY"*

On November 9, 1958, the *Lady Be Good* was located in the Libyan Desert by an oil exploration team. Broken in two, perfectly preserved by the dry desert air, and with no signs of the crew within miles of its crash site, the plane was quickly dubbed: *"The Ghost Bomber of WWII."* Eventually, the remains of eight of the nine crewmen were found 75–100 miles northwest of

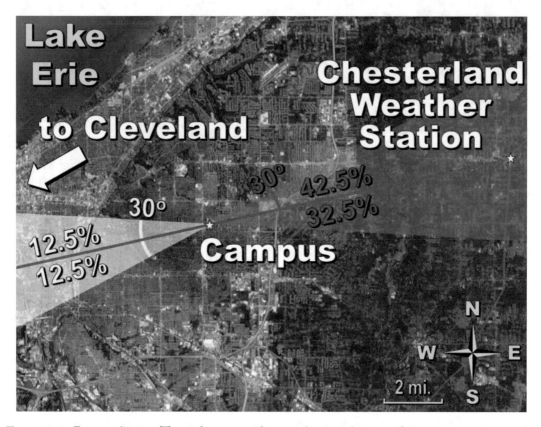

Figure 3.8: Regional map. The red arrow indicates the true bearing from our campus to the weather station. Results show that all students were able to pinpoint the correct line of transmission to within 15°. However, 75% of students chose a bearing facing the weather station (as indicated by the red area), while 25% of students fell victim to the *"180°-ambiguity"* and chose a bearing away from the weather station (as indicated by the yellow area).

Figure 3.9: Campus map. Small arrows indicate the bearing that each student determined. Red arrows face toward the weather station while yellow arrows face away from the station.

the wreck (see Figures 3.10, 3.11, and 3.12). Somehow, on the return leg of its mission, the crew had missed their home base and flown 440 miles southeast of Soluch, deep into Libya's Kufra district. Low on fuel, the crew parachuted to safety just before the plane crashed into the Calanshio Sand Sea. Diaries recovered from the desert indicate that eight of the nine men survived their jumps (one man's chute failed to open) but were so lost that they initially thought they were bailing-out into the Mediterranean Sea. When they landed on the desert floor, they mistakenly thought they were within walking distance of their base, but in reality, the plane was in the middle of the Sahara Desert. With only half of a canteen of water and unaware of their location, the survivors agonized in vain for eight days to walk in a northwest direction back to Soluch [10, 20, 21]. The crew may actually have survived had they instead chosen to walk south to the Oasis of Wadi Zighen. Five of the crew died after walking 75 miles while two crewmen managed to cover an astonishing 100 miles before succumbing to the desert. The remains of the eighth crewman were never found. The diaries of Co-Pilot, Second Lieutenant Robert Toner and Flight Engineer Technical Sergeant Harold Ripslinger chronicle the crew's steady decline into mental and physical exhaustion. Their daily entries are heartbreaking to read [22].

Subsequent analysis indicates that the final flight of the LBG was plagued by four contributing factors—chief among them was the inexperience of the crew. Only Hatton had been on a previous mission and that was as a Co-Pilot on an aborted bombing run only two days earlier. At 27 years, Toner was the oldest man in Hatton's crew. Next, the plane veered significantly off course during the flight from Soluch to Naples because of weather conditions and strong headwinds. The circuitous outbound flight meant that the plane would be returning alone, in the dark, and with a strong accompanying tailwind. Third, the crew jettisoned its 4,500-pounds of bombs over the Mediterranean and Hatton decided to return to base at a higher altitude than planned in an effort to improve his view. The higher altitude, lighter plane, and accompanying tailwind gave the plane a higher ground speed than anticipated. This placed the LBG southeast of the Benina Tower, not northwest, when it began searching for the RF signal to guide it home. Finally, the appearances of sand and sea at night is confusingly similar. Thinking they were over the Mediterranean Sea and inbound to North Africa, Pilot and Co-Pilot expected to see, and indeed saw, nothing but darkness below their plane. Low on fuel and thinking they were over water, the crew bailed out of the plane carrying provisions for survival at sea, taking life jackets instead of water and rations.

Despite these contributing factors, the official investigation report cites the primary cause of the plane's disappearance as the rookie navigator's misinterpretation of the RF directional bearing emitted by Benina Tower. Like our students, the navigator had fallen victim to the "180°-ambiguity." As the plane was heading back to Soluch, it had already passed the tower when Hatton radioed for an *inbound* bearing. However, a reading off of the RF front appears identical to a reading off of the RF back (called the *"reciprocal reading"*). When the navigator detected a bearing of 330°, he assumed the plane was still northwest of Benina Tower and that the plane was *inbound* to Soluch. In reality, the LBG was already southeast of Benina and the navigator

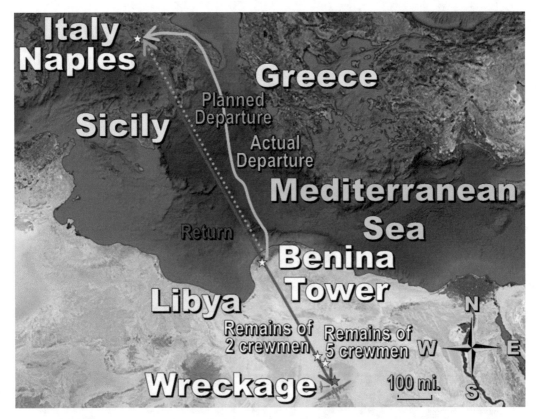

Figure 3.10: The actual flight of the *LBG*. Strong headwinds pushed the plane significantly off course during the outbound leg of its journey. After unloading its bombs in the Mediterranean Sea, the plane flew in a southeast direction toward Benina Tower. Strong tailwinds pushed the plane past Benina when the crew took a reading off of the back of the tower (called the *"reciprocal reading"*). Mistakenly thinking they were inbound to Soluch, the plane flew 440 miles into the Libyan Desert. Bearings off the front and back of Benina Tower would have been identical to the navigator of the LBG. The purple star indicates the location of the remains of Bombardier, Second Lieutenant John S. Woravka, who did not survive the bail-out. The remains of the Waist Gunner/Radio Operator, Staff Sergeant Vernon Moore were never found. Map courtesy of *Google Maps*.

Figure 3.11: The wreckage of the *Lady Be Good* as it was first seen by air.

Figure 3.12: Investigators from a search party explore the wreckage.

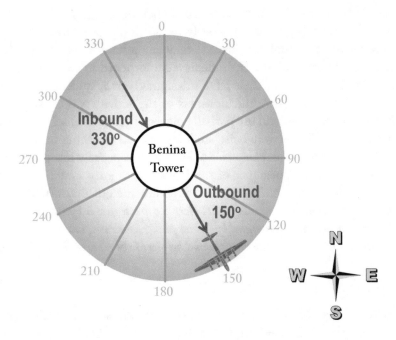

Figure 3.13: The *"180°-ambiguity."* When Hatton asked for an *inbound* emergency bearing, the plane had already passed Benina Tower. Thus, the reported bearing of *inbound* 330° was actually *outbound* 150°.

was reading the reciprocal off of the RF back. Tragically, the navigator's misinterpretation of the reciprocal reading meant that the LBG's *inbound* bearing of 330° *to* the tower was actually an *outbound* bearing of 150°(330° − 180° = 150°) *away from* the tower (see Figure 3.13)! [23]. The navigator became confused and directed the plane deeper and deeper into the desert, albeit along the correct flight path. **In short, the navigator made the same mistake that 1/4 of our students did!**

3.2 CONCLUSIONS

The HANDI-Finder® used in our activity differs significantly from the rotatable loop antenna used aboard the LBG, yet both systems are afflicted by the same inherent ambiguity. The goal of our activity was not in re-creating the exact navigational electronics of a WWII B-24 bomber, but in emphasizing to the students that despite the very latest in technological advances and instrumentation, judgments are still a necessary part of scientific endeavors. Our activity is as much a human experiment as it is a technical one. The fate of the LBG occurred because seemingly unrelated factors aligned to form a *"perfect storm"* that led to a catastrophic conclusion. Had any one of these factors been absent or avoided, the plane and its crew would have most likely survived its first combat mission [24]. As mentioned before, we did not tell students about the "180°-*ambiguity*" of our RDF's design. Obviously, results would have turned out quite differently if students were aware of the ambiguity because they might have sought additional information to discern the correct bearing. However, students were surprised to see how easily they fell victim to a somewhat obvious flaw in their RDF's design. Overall, students responded extremely favorably to our activity. The most difficult aspect of the activity was constructing the RDFs. To save time and headaches, we will use partially pre-built versions in subsequent implementations. However, once the RDFs were built, students enjoyed searching for the direction of the RF signal at different locations on campus. Students were especially intrigued by the historical context of the mysterious disappearance of the LBG and to learn how the laboratory activity connected to the plane's final fate. As one student commented in a post-activity survey, *"I now see how easy the navigational error was to make. Here we are almost 80 years later making the exact same mistake … and we weren't surrounded by fighter planes, darkness, and unfamiliar surroundings. This gave me a true appreciation for the bravery and heroics of the crew of the Lady Be Good."*

3.3   ACKNOWLEDGMENTS

The authors acknowledge Richard Rarick, for his usual proofreading and editorial commentary, as well as the reviewers of this manuscript. The work of these individuals significantly strengthened the presentation of this work.

3.4   REFERENCES

[1] Photograph courtesy of the U.S. Air Force. 37

[2] G. A. DiLisi, A. Chaney, K. Kane, and R. Leskovec, The lady be good: A case study in radio frequency direction finders, accepted by *Phys. Teach.*, July 2020. 38

[3] G. A. DiLisi and R. A. Rarick, Monday night football: Physics decides controversial call, *Phys. Teach.*, 41:454–459, November 2003. DOI: 10.1119/1.1625203. 38

[4] G. A. DiLisi and R. A. Rarick, Modeling the 2004 tsunami for introductory physics courses, *Phys. Teach.*, 44:585–588, December 2006. 38

[5] G. A. DiLisi and R. A. Rarick, Bouncing back from "Deflategate," *Phys. Teach.*, 53:341–346, September 2015. DOI: 10.1119/1.4928347. 38

[6] G. A. DiLisi and R. A. Rarick, Remembering the S.S. Edmund Fitzgerald, *Phys. Teach.*, 53:521–525, December 2015. DOI: 10.1119/1.4935760. 38

[7] G. A. DiLisi, The Hindenburg disaster: Combining physics and history in the laboratory, *Phys. Teach.*, 55:268–273, May 2017. DOI: 10.1119/1.4981031. 38

[8] G. A. DiLisi and S. McLean, The Apollo1 fire: A case study in the flammability of fabrics, *Phys. Teach.*, 57:236–239, April 2019. DOI: 10.1119/1.5095379. 38

[9] G. A. DiLisi, A. Chaney, S. McLean, and R. A. Rarick, A case studies approach to teaching introductory physics, *Phys. Teach.*, 58:156–159, March 2020. DOI: 10.1119/1.5145402. 38

[10] D. E. McClendon, *The Lady Be Good: Mystery Bomber of World War II*, Aero Publishers, Inc., Fallbrook, CA, 1962. 41, 50

[11] R. A. Leskovec, *The HANDI-Finder® Experimenter's Kit*, RALTEC® Electronics, Highland Heights, Ohio, 2017. 43

[12] R. A. Leskovec, Build the HANDI-Finder!, *QST*, pages 35–38, May 1993. 43

[13] https://www.rainbowkits.com/kits/sdf-1p.html 45

[14] https://www.handi-finder.com/ 45

[15] Information on the atan2 function can be found at functions.wolfram.com. Retrieved from http://functions.wolfram.com/PDF/ArcTan2.pdf on February 4, 2020. 46

[16] Note that the *Microsoft Excel* reverses the arguments of its atan2 function. 46

[17] R. Johnson, Spherical trigonometry, *West Hills Institute of Mathematics*, February 4, 2020. https://www.math.ucla.edu/~robjohn/math/spheretrig.pdf DOI: 10.2307/2302404. 46

[18] http://mathworld.wolfram.com/SphericalTrigonometry.html 46

[19] https://www.movable-type.co.uk/scripts/latlong.html or https://www.igismap.com/formula-to-find-bearing-or-heading-angle-between-two-points-latitude-longitude/ 46

[20] R. Barker, *Great Mysteries of the Air*, The Macmillan Company, New York, 1967. 50

[21] M. Martinez, *Lady's Men: The Story of World War II's Mystery Bomber and Her Crew*, The U.S. Naval Institute Press, Annapolis, MD, 1994. 50, 55

[22] Diary of Robert Toner, July 27, 2018. http://www.ladybegood.net/diaries/ 50

[23] Author Mario Martinez claims that the fate of the LBG is shared by another radio operator at nearby Benghazi who failed to respond to Hatton's plea for a position report, believing the airplane to be German. See reference [21]. 53

[24] For an excellent analysis of the LBD's final voyage, see the show Ghost bomber: The Lady Be Good, July 27, 2018. https://www.youtube.com/watch?v=d9W5P3-mxwY 54

CHAPTER 4

Using Operational Definitions

Case studies highlight the importance of operational definitions in scientific experiments

With a case study, students see, perhaps for the first time, how operational definitions compare phenomena of interest against known standards or accepted protocols. Our article on the *Hindenburg* disaster provides an excellent case study to illustrate this point. On May 6, 1937, the German zeppelin *Hindenburg* caught fire while preparing to dock at the Naval Air Station in Lakehurst, NJ (see Figure 4.1). The ensuing fire destroyed the massive airship in 35 seconds. We present the historical debate as: *"What was the source of fuel for the fire that destroyed the Hindenburg?"* In the laboratory, we conduct a vertical flame test that compliments the horizontal flame test of Chapter 5. The vertical test is patterned after the procedure used by the American Society for Testing and Materials (ASTM) for determining the flammability of textiles. Using the ASTM test demonstrates the importance of operational definitions in scientific experiments. The purpose of the flame test is to not only raise the historical awareness of our students, but to acclimate them to the idea of adhering to standardized scientific protocols.

4.1 THE HINDENBURG DISASTER: COMBINING PHYSICS AND HISTORY IN THE LABORATORY, A CASE STUDY IN THE FLAMMABILITY OF FABRICS (VERTICAL FLAME TESTS)

Gregory A. DiLisi, *John Carroll University, University Heights, Ohio*

4.1.1 INTRODUCTION

On May 6, 1937, the German passenger zeppelin, ***Hindenburg*** [1], hovering 300 feet in the air and held aloft by 7 million cubic feet of hydrogen gas, burst into flames while preparing to dock at the Naval Air Station in Lakehurst, NJ (see Figure 4.1). Amazingly, the ensuing fire consumed the massive airship in only 35 seconds! In the aftermath, 35 of 97 people onboard died (13 passengers and 22 crewmen) plus 1 member of the ground crew. Herbert Morrison, the broadcaster from Chicago's WLS radio station, was on assignment that day covering the arrival of the majestic airship. Morrison's eye-witness account of the disaster is legendary audio history. In fact, Morrison's phrase, "Oh, the humanity!" has become a cultural idiom.

Figure 4.1: The Hindenburg disaster, May 6, 1937. (Public Domain.)

We present the Hindenburg disaster as a case study in the flammability of fabrics. Our goal is to examine the ship's outer covering and decide whether or not it was the fire's initial source of fuel. To accomplish this, we piloted a basic vertical flame test with students in an introductory-level undergraduate laboratory. Our test is patterned after the protocol set forth by the American Society for Testing and Materials (ASTM) for determining the flammability of textiles. The case study provides several unique teaching opportunities: mainly, how using an ASTM protocol for determining the flammability of textiles demonstrates the importance of operational definitions in scientific experiments.

4.1.2 "THIS GREAT FLOATING PALACE"

Well, here it comes, ladies and gentlemen … No wonder this great floating palace can travel through the air at such a speed, with these powerful motors behind it … The sun is striking the windows of the observation deck on the eastward side and sparkling like glittering jewels on the background of black velvet [2].

The zeppelin, LZ 129 Hindenburg (Luftschiff Zeppelin #129), was launched on May 4, 1936 as the premier passenger aircraft of the world's first airline, "*Deutsche Luftschiffahrts-Aktiengesellschaft*" *(DELAG)*, or the "German Airship Transportation Corporation." Named in honor of the late President of Germany, Paul von Hindenburg, the ship was the largest object ever to fly, stretching an incredible 803.8 ft in length and boasting a maximum diameter of 135.1 ft. In comparison, the Titanic was 883 ft in length while a modern Boeing 747 is only 250 ft. A double-decked gondola, constructed mainly inside the hull to reduce aerodynamic drag, provided passengers with unparalleled luxury: 72 sleeping berths, washrooms, dining room and bar, pressurized smoking room, and spacious lounge. A promenade with slanted windows allowed passengers to gaze upon the scenery below. With four reversible 1,200-hp diesel engines and cruising speed of 76 mph, the airship provided the quickest means of crossing the Atlantic, cutting the typical transit time of contemporary ocean liners in half [3].

The ship was classified as a rigid airship because of its steel frame. Within the steel structure were 16 large gas cells (or bladders) made of gelatinized latex, designed originally to hold inert *helium* gas. However, at that time, only the United States, which had stockpiled the non-flammable gas as a byproduct of its mining of natural gas, had enough helium to supply a fleet of airships. Even though the U.S. Helium Control Act of 1927 prohibited American export of helium to any foreign nation, DELAG was hopeful it could convince the U.S. government to export it. Unfortunately, tensions between the U.S. and Germany deteriorated so the export ban was never lifted. Therefore, DELAG made the decision to trade safety for cost and re-engineered the bladders to hold seven million cubic feet of *hydrogen* as the lifting gas. Converting to hydrogen also had the added benefit of giving the Hindenburg more lift, increasing its capacity to an impressive 242.2 tons of gross lift and 112.1 tons of useful lift. From the ship's control room, the crew could drop water ballast or release hydrogen gas from ventilation shafts along the top of the ship to adjust its buoyancy and trim. Last, the steel structure was covered by panels of cotton cloth doped with various compounds. These panels were stitched together to form a single, and presumably electrically continuous, "skin." The flammability of this outer covering plays a pivotal role in the current debate surrounding the ship's destruction so we postpone a detailed description of the fabric until later in this chapter. By May 1937, the Hindenburg was making the first flight of its second season of service, having already completed several safe trans-Atlantic journeys in 1936. Public confidence in hydrogen-filled airships was soaring.

4.1.3 "THE SHIP IS RIDING MAJESTICALLY TOWARD US"

The ship is riding majestically toward us like some great feather ... It's practically standing still now; they've dropped ropes out of the nose of the ship, and (uh) they've been taken ahold of down on the field by a number of men. It's starting to rain again.

On what would be its final voyage, the Hindenburg left Frankfurt, Germany at 7:16 pm on May 3, 1937, under the command of Captain Max Pruss and First Officer Albert Sammt. This was

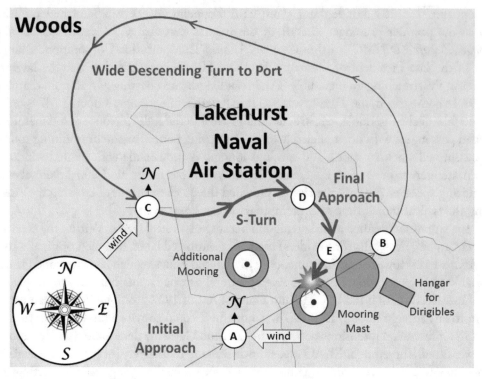

Figure 4.2: Approach of the Hindenburg. The thickening red line represents a lowering elevation of the ship.

Pruss's first time commanding the Hindenburg. The ship was scheduled to arrive at Lakehurst, NJ at 6:00 am on May 6 but unusually strong headwinds caused it to run several hours behind schedule. As the ship approached New Jersey, it encountered a storm before reaching the Lakehurst Naval Air Station at 4:15 pm. The airfield was under the command of Charles Rosendahl who radioed the ship to delay landing until weather conditions improved. By 6:22 pm, the storm had passed but conditions were rapidly worsening. Rosendahl radioed the Hindenburg, recommending "the earliest possible landing."

To understand the Hindenburg's final approach to the Lakehurst Naval Station, note that wind direction refers to the direction from which the wind is coming. Also, navigational directions are denoted as follows: "bow" refers to the front of the ship; "stern" refers to the back of the ship; "port" refers to the left side of the ship (when facing forward); and "starboard" refers to the right side of the ship (when facing forward). Facing an easterly wind, Pruss approached the airfield at 7:08 pm from the southwest at an elevation of about 650 ft to observe ground conditions (Figure 4.2—Point A). At 7:09 pm, he initiated a wide descending turn to port in order to dock the ship pointing into the wind. This would maximize the ship's aerodynamic stability as passengers disembarked (Figure 4.2—Point B). Sammt lowered the Hindenburg's elevation by releasing hydrogen in 15-second intervals from various ventilation shafts as the ship turned. At 7:16 pm, Pruss was lining the ship to the east at an elevation of about 400 ft when the wind shifted from easterly to southwesterly, forcing Pruss to again turn the ship for docking, this time facing southwest (Figure 4.2—Point C). With deteriorating weather conditions, little room to maneuver, and anxious to land, Pruss decided to execute a tight S-turn rather than make another large looping pass over the airfield. He ordered a sharp turn to port followed by a sharp turn to starboard. Sammt continued to vent hydrogen from various gas cells to lower the ship's elevation to 350 ft. At some point prior to, or during this S-turn, the ship began to run heavy in the tail because at approximately 7:18 pm, Sammt ordered 2 drops of 300 kg of water ballast from the stern and valved 5 seconds of hydrogen from the bow (Figure 4.2—Point D). At 7:19 pm, with the tail still heavy, Sammt ordered one last drop of 500 kg of water ballast from the stern and sent six crewmen to the front of the ship to help lower the bow. Finally, at 7:21 pm, the ship was at an elevation of 300 ft and roughly pointed into the wind (Figure 4.2—Point E). Although the ship was still heavy in the tail, the forward grounding lines were dropped. A light rain began to fall. The metal frame was now electrically grounded by the landing lines.

4.1.4 "IT'S BURST INTO FLAMES"

It's burst into flames! … and it's crashing! It's crashing terrible! Oh, my! … It's smoke, and it's in flames now; and the frame is crashing to the ground, not quite to the mooring mast. Oh, the humanity! And all the passengers screaming around here … Listen, folks; I … I'm gonna have to stop for a minute because I've lost my voice. This is the worst thing I've ever witnessed.

Using eyewitness accounts to determine the origin of the fire proved confusing but the first sign of trouble appears to have been at the top, rear of the ship, just in front of the vertical fin [4]. Both R. H. Ward (stationed with the port bow landing party) and R. W. Antrim (stationed atop the mooring mast) testified that they noticed a fluttering of the ship's outer cover at this location— suggesting hydrogen was leaking out of a rear interior bladder against the outer covering [4]. Crewmen in the control stations of the lower fins testified hearing "muffled detonations" near the top of the ship. When they looked up, they saw bright red and yellow reflections of fire [4]. By 7:25 pm, a yellow flame appeared on the outside of the ship at this spot. Within seconds, the tail section was engulfed in flames. The ship managed to stay afloat for a few seconds but eventually, the tail section sank, slamming crew and passengers 15–20 ft backward into the rear walls of the control room, cabins, dining lounge, and promenade. As the Hindenburg tilted upward, the fire traveled inside the ship along the central axis until a blowtorch of fire erupted from the nose (Figure 4.3). Crewmen stationed in the bow were incinerated.

Most eye-witnesses described the Hindenburg as burning from the inside-out. Within 30 seconds, the entire ship crashed to the ground and rolled slightly starboard. In general, passengers and crew in the promenade or public areas near the outside of the ship were able to jump

Figure 4.3: The final moments of the Hindenburg. As the bow angled upward, a blowtorch of fire erupted out of the nose. (Public Domain.)

to safety while those deeper inside the ship (interior cabins and control stations) did not. Fortunately, many passengers had gathered in the promenade to watch the landing. Some family members lived or died based merely on a few feet of separation. Pruss and Sammt stayed with the ship until it hit the ground. Both men survived the crash but Pruss was badly disfigured from burns he received carrying crewmen from the wreckage. On the ground, Herb Morrison had been assigned to cover the landing because of his prior work in broadcasting from an airplane. Normally, he would have been in Chicago covering a live musical program. After the tragedy, the 16-inch green lacquer disk recordings of Morrison's account, which were actually damaged by debris from the burning airship, were flown by airplane to Chicago and broadcast that night from radio station WLS. In 1987, a small memorial pad was dedicated on the 50th anniversary of the tragedy. The pad is located on the site where the Hindenburg's gondola landed.

4.1.5 THEORIES

Film footage coupled with Morrison's audio account brought a swift end to the era of zeppelins. The public would never again feel safe aboard one. Subsequent investigations by the U.S. and

Germany were inconclusive in determining the cause of the fire. For years, scientists, politicians, and military personnel put forth several theories as to the underlying causes of the disaster. Was it sabotage? No evidence of sabotage was ever found. Was it a lightning strike? Unlikely—the outer covering of the ship had several burn-holes, some as large at five centimeters in diameter, proving the ship had survived in-flight lightning strikes during its first year of service [5]. Today, a re-examination of the evidence leaves us with two competing theories that at least agree on the fire's *source of ignition*. As the Hindenburg passed through the storm off the New Jersey coast, it became electrically charged. When the landing lines touched the ground prior to docking, they "earthed" the Hindenburg's steel frame, but not every panel of the ship's fabric covering. A spark between the charged panel of fabric and the grounded steel frame ignited some source of fuel. The difference between the two theories lies in identifying that *source of fuel.*

- **Leaking hydrogen gas:** The most likely explanation of events is that the electrostatic discharge ignited leaking hydrogen gas. Recall that during the sharp S-turn, Sammt was unable to correct the ship's trim. Experts agree that the ship was undoubtedly leaking hydrogen from the stern [4]. What caused the leak? One theory suggests that the S-turn was uncommonly tight and that one of the rudder's bracing cables may have been over-stressed to the point where it snapped and slashed through a gas cell [6]. Maybe something as simple as a sticky valve was at fault. Regardless of the cause of the leak, an explosive mixture of hydrogen gas and air floated above the ship's tail. Because this theory blames the ship's demise on the decision to land after the storm and on the S-turn made just prior to docking, it was strongly refuted by Pruss and Rosendahl, both of whom always maintained the ship fell victim to sabotage.

- **Incendiary paint:** In 1997, engineer Addison Bain put forth the idea that at least early in the fire, the ship's outer covering itself, and not leaking hydrogen gas, was the primary source of fuel for the fire [7]. The cotton cloth that covered the ship was doped with different mixtures based on cellulose acetate butyrate (CAB), the base resin for what are commonly called "lacquers." The portion of cloth covering the lower half of the ship was doped with a layer of pure CAB then three layers of CAB mixed with aluminum powder. The portion of cloth covering the upper half of the ship was doped with a layer of pure CAB, a layer of CAB mixed with iron oxide, and three layers of CAB mixed with aluminum powder. These coatings gave the ship its distinctive reflective appearance and were used to keep the outer skin taut for aerodynamic purposes as well as to protect it from wind, water, and small objects. The added layer of iron oxide on the upper portion of the ship protected the interior gas bags from damage by UV radiation and overheating from IR radiation. Bain developed his theory when he realized that these compounds are *similar* to the components of thermite, a pyrotechnic composition that resembles a common sparkler or the propellant in the Space Shuttle's Solid Rocket Boosters. In short, Bain argued that the Hindenburg's outer skin was essentially a gigantic sparkler. According to Bain, the electric discharge was energetic enough to ignite the skin and cause a dramatic exothermic reduction-oxidation reaction—therefore, this idea has become known as the "Incendiary Paint Theory" (IPT). The IPT has merit for two reasons: (i) Hydrogen burns with an invisible flame yet the Hindenburg was consumed in an enormous yellow and red fireball. One might conclude that something other than hydrogen was burning. (ii) The ship held its position for a few seconds before the stern crashed to the ground. One might conclude that the gas cells were intact when the fire started. To test his idea, Bain obtained an actual remnant of fabric from the Hindenburg and ignited it with a continuous spark. The piece burned as a brilliant yellow burst that looked like a miniature version of the Hindenburg disaster [8].

4.1.6   STATEMENT OF THE PROBLEM

We used the IPT as the basis of a new, inexpensive lab activity focusing on a previously-untapped topic in our course, namely the flammability of fabrics. Since the IPT posits that the propagation of the fire was due to burning of the Hindenburg's fabric, we designed an investigation to quantify how fabrics burn after the source of ignition is removed. Our activity is modeled after the vertical flame test, ASTM D 6413-99—*The Standard Test Method for Flame Resistance of Textiles (Vertical Test)*, which has been adopted as an accepted Federal Test Standard [9]. This protocol is considered to be the most fundamental and commonly used test on flame resistant fabrics in the U.S. To align with the ASTM protocols, English units are used throughout the analysis. The equipment and materials needed to create this activity cost under $30 and were readily obtained from local hardware, fabric, and automotive refinishing stores. We piloted our activity on a cohort of undergraduate students. We started by presenting students with the historical background information described in the previous sections. Next, we emphasized two concepts: (i) the goal of our activity is not to prove nor disprove the IPT, but to showcase how physics can be used in the real world; and (ii) even though our activity focuses only on a *vertical* flame test, it gives us a quantitative understanding of how flammability is tested and how results can be used to unravel the Hindenburg disaster (obviously, combining our results with those from a *horizontal* flame test would provide a more complete analysis).

Sample preparation: Because purchasing chemicals is expensive and heavily regulated, reproducing the Hindenburg's outer fabrics is simply not possible for most teachers. Instead, we chose to test three fabrics that were easy-to-make, inexpensive to buy, yet nicely approximate the outer coverings of the airship. First, we purchased swatches of pure cotton cloth and ironed them. Next, we purchased clear lacquer, black primer with iron oxide as its tinting agent (iron oxide black pigment, in powder form, can also be purchased online), and a concentrated aluminum resin paste (our specific type was Genesis LV 1060 from Sherwin–Williams Paints). To create the fabric that approximates the covering on the upper portion of the ship, we rolled cotton swatches with a layer of clear lacquer, then a layer of black primer, then three layers of aluminum paste. To create the fabric that approximates the covering on the lower portion of the ship, we

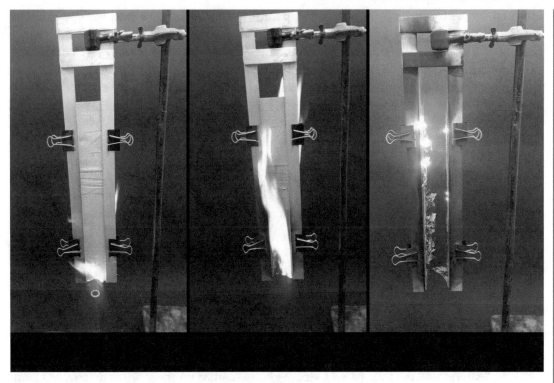

Figure 4.4: Our apparatus for vertical flame testing.

rolled cotton swatches with a layer of clear lacquer then three layers of aluminum paste. After the coatings dried, we were left with samples of pure cotton cloth and two stiff, reflective fabrics that approximate the fabric on the lower and upper portions of the Hindenburg. Students trimmed each of these three samples into five 12-inch × 3-inch strips and weighed each strip. Each strip was placed into a frame of sheet metal that secured the strip on two sides, leaving the bottom edge exposed. The frame was clamped together at four locations and suspended in a laboratory hood.

Testing: A flat black poster background and dimmed lights were used to enhance observation of burning fabrics inside the hood. Precautions were taken to minimize drafts in the hood. No attempt was made to control ambient temperature or pressure. No attempt was made to move the strips nor to test a horizontal orientation of the strips. Each strip was tested and the average of five strips was reported per fabric. A Bunsen burner, with 10-mm inside diameter barrel, was used to create a 1.5-inch high, 99%-pure methane, flame. The burner could be swiveled so that the exposed edge of the strip was exactly 0.75 inches above the top of the burner. The flame was applied for 12 ± 0.25 seconds (flame-to-strip), as measured by a stopwatch. Students filmed each trial using cell phone cameras in "slow motion" mode (Figure 4.4a). Once the flame was removed, students continued to film the strip until any visual flame or glow self-extinguished (Figure 4.4b and Figure 4.4c). These videos would be used only to determine the duration of time, to 0.1 second resolution, that the samples burned and/or glowed. Any signs of melting or dripping were noted. If a portion of the strip remained intact, the strip was removed from the metal frame so that students could apply the following specific tearing force to the strip: a crease was made running lengthwise through the peak of the highest charred area and parallel to the side of the strip. A hook was inserted into the strip 0.25 inches from the charred edge. A weight was attached to the hook depending on the strip's weight per unit area (100 g for strips 68–203 g/m^2; 200 g for strips 204–508 g/m^2; 300 g for strips 509–780 g/m^2; and 400 g for strips over 780 g/m^2). With the hook in place, students grabbed the other side of the charred edge and raised the strip in a smooth continuous motion until the tearing of the strip along the crease stopped. Many strips were totally consumed by the vertical flame so tearing the strip was not necessary (Figure 4.4c).

4.1.7   ANALYSIS

Using their video-clips and a ruler, students determined or calculated the following.

- **"Afterflame,"** the time when a visible flame remained on the strip, as determined from the cell phone video clips.

- **"Afterglow,"** the time when a visible glow remained on the strip, as determined from the cell phone video clips.

- **"Char Length,"** the distance from the exposed edge of the strip to the furthest point of visible damage after the tearing force was applied, as determined by measuring the damaged portion of the strip with a ruler. Note that the Char Length is 12 inches if the strip is totally consumed by the fire.

- **"Vertical Burn Rate,"** calculated as the Char Length divided by the Afterflame.

- **"Vertical Burn Time,"** (designated as t_{lower} or t_{upper}), the time, extrapolated from the vertical burn rate, needed to burn a 106.1-ft long swatch of the strip, (i.e., 1/4 of the Hindenburg's maximum circumference). Because the upper and lower portions of the ship were covered in different fabrics, the maximum vertical distance a fabric could burn is $\sim$ 1/4 the maximum circumference of the ship (see Figure 4.5).

- **"Total Vertical Burn Time,"** the total time needed to burn vertically one entire side of the ship. This time was determined by adding the vertical burn times for the fabrics covering the lower (t_{lower}) and upper (t_{upper}) portions of the ship.

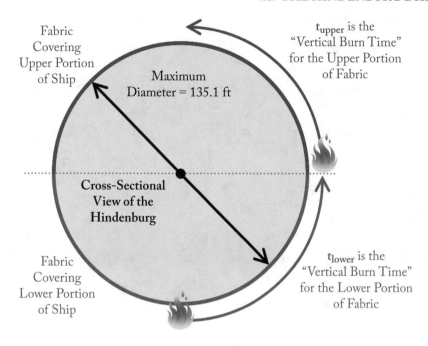

Figure 4.5: Depiction of the vertical burn time.

4.1.8   RESULTS

Table 4.1 shows data from a typical run of our activity. Results support the notion that leaking hydrogen, and not incendiary paint, is the most plausible source of fuel for the fire that consumed the Hindenburg. The outer fabrics just do not burn at a fast enough rate to consume a ship the size of the Hindenburg in a timeframe of the order of a minute. After burning strips of pure cotton cloth, students dramatically see that the dopants used on the Hindenburg's fabric actually retard the spreading of fire. The dataset below shows that a fire would need ~ 112 minutes to burn a distance roughly equal to the *height* of the ship, that is to say, from the underside to the topside of the ship. Experiments conducted by A. J. Dessler on *horizontal* burn rates support our results and show a fire would need 11–12 hours to burn a distance roughly equal to the *length* of the ship [10, 11]. Dessler discounts the two merits of the IPT explaining that the yellow fireball was actually the fabric, wires, and steel girders burning with visible flames in the invisibly burning hydrogen (as a mantle burns visibly in a lantern even though the gas that is actually burning may be invisible). Also, as the hydrogen burned the outer fabric and rushed out of the bladders, air rushed in. This created updrafts at the tail of the ship that were strong enough to keep the ship momentarily afloat—just as a burning piece of paper is lifted by the updraft created by its own fire. Finally, a 2007 episode of the popular show, "Myth Busters" (Episode 70—"The Hindenburg Mystery"), found similar results and suggested that the IPT, at

Table 4.1: Typical run of activity

Weight per Area (g/m²)	Afterflame (sec)	Afterglow (sec)	Char Length (inches)	Dripping or Melting	Vertical Burn Rate (in./sec)	Vertical Burn Time [extrapolated] (min)
Cotton Cloth						
114.8	4.1	5.5	12	No	2.9	7.3
Approximation of fabric covering lower portion of ship (Cotton cloth + 1 layer of lacquer + 3 layers of aluminum paste)						
287.0	30.1	32.4	12	No	0.40	t_{lower} = 53.2
Approximation of fabric covering upper portion of ship (Cotton cloth + 1 layer of lacquer + 1 layer of black primer + 3 layers of aluminum paste)						
344.4	33.4	36.8	12	No	0.36	t_{upper} = 59.1
Total Vertical Burn Time = t_{lower} + t_{upper}						112.3 min

least in the open sources, is doubtful [12]. Although the episode is not peer-reviewed and should be viewed with some skepticism, it is a phenomenal visual resource available to any teacher and can be shown to students to emphasize or solidify certain concepts. The episode serves as a good closure activity and can be downloaded from iTunes for $1.99.

4.2 CONCLUSIONS

The 80th anniversary of the Hindenburg disaster presents a compelling case study that brings powerful teaching opportunities to a variety of disciplines. First, the anniversary raises historical awareness in our students while bringing real-world applications of physics to them. Next, the physics of flammability can be treated appropriately at the introductory level since only careful measurements of time, distance, and weight are needed. In fact, our case study can serve as a start-of-the-semester laboratory exercise where safety, measurement, and error analysis are emphasized. Conversely, our case study can also serve as a capstone project in a senior-level engineering course after which each student is required to examine another engineering failure from history (e.g., Chernobyl, Three Mile Island, the collapse of the Tacoma Narrows Bridge, the Challenger disaster, etc.) and design an experiment with relevant operational definitions to test proposed hypotheses. Third, the resulting analysis shows students that just as scientific theories are open to re-examination in the light of new or confounding observations, so too are historical events open to re-visitation and scrutiny. Next, the case study demonstrates to students how a cascade of unlikely events can result in an unpredictable catastrophe and how scientists and engineers often test proxies when actual materials are unavailable or prohibitive to examine directly. Finally, our case study perhaps provides the first opportunity to introduce students to testing standards and how the responses of materials and products are determined.

4.3   ACKNOWLEDGMENTS

The author acknowledges James DeLuca, chemist extraordinaire, for his insights into sample preparations. The author also acknowledges the reviewers of this manuscript who significantly strengthened the presentation of this work.

4.4   REFERENCES

[1] G. A. DiLisi, The Hindenburg disaster: Combining physics and history in the laboratory, *Phys. Teach.*, 55:268–273, May 2017. DOI: 10.1119/1.4981031. 57

[2] H. Morrison, WLS Radio (Chicago), Address on the arrival of the airship Hindenburg [transcription of audio file], Lakehurst, NJ. (May 6, 1937). http://www.americanrhetoric.com/speeches/hindenburgcrash.htm on January 3, 2017. 58

[3] B. Waibel, *The Zeppelin Airship LZ:129 Hindenburg*, pages 7–22, L.E.G.O.S.p.A., Italy, 2013. 59

[4] *Bureau of Air Commerce, Air Commerce Bulletin of August 15, 1937*, 9(2):28–29, United States Department of Commerce, Washington, DC, 1937. 61, 63

[5] R. Archbold, *Hindenburg: An Illustrated History*, Warner Books, New York, 1994. 63

[6] T. Graham, Hindenburg: Formula for disaster, *ChemMatters*, December 2007. 63

[7] A. Bain, Colorless, nonradiant, blameless: A Hindenburg disaster study, *Gasbag Journal/Aerostation*, 39, March 1999. 63

[8] Kurdistan Planetarium, *What Happened to the Hindenburg?*, January 3, 2017. https://www.youtube.com/watch?v=phZI_h3Q4V0 64

[9] American Society for Testing and Materials, *ASTM D:6413–99, Annual Book of ASTM Standards, Standard Test Method for Flame Resistance of Textiles (Vertical Test)*, ASTM, West Conshohocken, PA, 2016. 64

[10] A. J. Dessler, D. E. Owens, and W. H. Appleby, The *Hindenburg* fire: Hydrogen or incendiary paint?, *Buoyant Flight*, 52(2–3), Jan/Feb and Mar/April 2005. 67

[11] A. J. Dessler, The Hindenburg hydrogen fire: Fatal flaws in the Addison Bain incendiary paint theory, *Lunar and Planetary Laboratory*, University of Arizona, Tucson, AZ, June 3, 2004. http://spot.colorado.edu/~dziadeck/zf/LZ129fire.pdf on January 3, 2017. 67

[12] MythBusters, *Hindenburg MiniMyth.*, January 3, 2017. http://www.discovery.com/tv-shows/mythbusters/videos/hindenburg-minimyth/ 68

CHAPTER 5

Demonstrating the Phenomenon of "Normalization of Deviance"

Case studies demonstrate the phenomenon of "Normalization of Deviance" that plagued several notorious engineering disasters

Often, groups of scientists and engineers go to extreme lengths to test and re-test various design concepts and construction techniques. Immersed in an isolated and high-stress environment, these groups develop a false sense that their products are "fail-safe," "full-proof," and "invincible." Over time, the group accepts a lower and lower standard of performance until that standard becomes the new "norm." This phenomenon is known as "Normalization of Deviance," since the departure from a higher, more robust standard has been normalized. This concept was first developed by Diane Vaughan during her investigation of NASA's *Challenger* disaster. In her study, Vaughan realized that the O-rings inside the Space Shuttle Transportation System's Solid Rocket Boosters had underperformed during the 24 flights prior to the *Challenger* mission. However, with each successful flight, the standards were lowered and lowered because the O-rings were "getting the job done." The lowering of standards simply became easier to accept. The sentiment at NASA was that one day, when budgetary restrictions and deadlines were relaxed, engineers would go back to using the high standards originally expected of the O-rings. Usually, when the phenomenon of "Normalization of Deviance" sets in, only a catastrophe causes a re-examination of standards. For this topic, a case study was designed to honor the crew of Apollo 1 (see Figure 5.1). On January 27, 1967, a fire swept through the interior of NASA's *"AS-204"* Command Module and killed American astronauts Chaffee, Grissom, and White during a rehearsal of their upcoming space flight. We present the historical debate as: *"What was the source of fuel for the Apollo 1 fire?"*

Figure 5.1: The *Apollo I* Mission Patch.

5.1 THE APOLLO I FIRE: A CASE STUDY IN THE FLAMMABILITY OF FABRICS (HORIZONTAL FLAME TEST)

Gregory A. DiLisi, *John Carroll University, University Heights, Ohio*
Stella McLean, *John Carroll University, University Heights, Ohio*

5.1.1 INTRODUCTION

On January 27, 1967, the interior of NASA's *"AS-204"* Command Module (CM), occupied by American astronauts Chaffee, Grissom, and White, caught fire during a rehearsal of its scheduled February 21 launch [1]. By the time the ground crew was able to open the hatch, the three astronauts had perished. On April 24, 1967, NASA announced that the flight would be officially re-designated, *"Apollo 1."* In this case study, we conduct a basic *horizontal* flame test, patterned after the protocols set forth by the Environmental Protection Agency (EPA) to measure the ignitability of solids. The laboratory activity is a complimentary exercise to the *vertical* flame test described in our previous article that examined the initial source of fuel for the fire that destroyed the massive German zeppelin *Hindenburg*, in 1937 [2]. Combining techniques from both case studies gives students a quantitative understanding of how the flammability of materials is tested and how a forensics approach to physics can be used to understand significant historical events.

Figure 5.2: The primary crew of *Apollo 1*. Left-to-Right: Edward H. White, II (Lt. Col., USAF—Senior Pilot), Virgil I. *"Gus"* Grissom (Lt. Col., USAF—Command Pilot), and Roger B. Chaffee (Lt. Cdr., USN—Pilot). (Courtesy of NASA.)

5.1.2   THE CREW

Commanding the first manned Apollo mission was 40-year old veteran astronaut Virgil I. Grissom, Lieutenant Colonel, USAF. (See Figure 5.2.) A native of Mitchell, Indiana and 1950 graduate of Purdue University's Mechanical Engineering program, Grissom was one of NASA's original class of astronauts, the famed *"Mercury 7 Astronauts."* Preferring to be called *"Gus,"* Grissom piloted the second suborbital Mercury flight on July 21, 1961, spending 15 minutes, 30 seconds in space. When his capsule, *"Liberty Bell 7,"* impacted the Atlantic Ocean on splashdown, its hatch mysteriously blew off. Grissom later argued that the capsule's emergency explosive bolts had malfunctioned. Water rushed into the capsule and Grissom narrowly escaped drowning. Liberty Bell was not as fortunate—it sank, sitting at the bottom of the Atlantic Ocean until it was recovered in 1999. Grissom's second flight was the first of the Gemini program. He and John Young orbited the earth three times in their capsule, nicknamed *The Unsinkable Molly Brown*" by Grissom. As the Command Pilot aboard Apollo 1, Grissom was seated in the left-most seat (facing the cockpit dash) and had access to the emergency cabin pressure relief valve, to be opened in the event of a cabin fire. Many NASA officials believe that had Grissom not died aboard Apollo 1, he would have been the first man to walk on the Moon. Chief of the Astronaut Office and fellow Mercury 7 Astronaut, Deke Slayton, was in charge of setting the astronaut rotation for flights. Slayton expressed his hope that one of the original astronauts would have been chosen for the honor: *"My first choice would have been Gus"* [3]. Gus Grissom was survived by his wife Betty and sons, Scott and Mark.

The Senior Pilot aboard Apollo 1 was another space veteran, 36-year old Edward H. White, II, Lieutenant Colonel, USAF. Born in San Antonio, Texas, White graduated from West Point and missed making the U.S. Olympic team in the 400-meter hurdles by a tenth of a second. After earning an M.S. degree in Aeronautical Engineering from the University of Michigan, he was selected by NASA in 1962 as part of the Group 2 class of astronauts. White flew aboard Gemini 4 and on June 3, 1965, became the first American astronaut to perform a spacewalk. White's EVA made him an instant and international celebrity. Images of White performing his iconic walk are as breathtaking today as they were in 1965. In fact, a depiction of White floating in space while tethered to his Gemini capsule was used on a pair of 1967 5¢ U.S. postage stamps. As Senior Pilot aboard Apollo 1, White was seated in the center and was responsible for opening the main hatch if the crew had to make an emergency egress from the cockpit. Ed White was survived by his wife Patricia and children, Edward III and Bonnie Lynn.

Rounding out the crew of Apollo 1 was 31-year old, Grand Rapids, Michigan native and Pilot, Roger B. Chaffee, Lieutenant Commander, USN. Like Grissom, Chaffee was a graduate of Purdue University, earning his Bachelors of Aeronautical Engineering in 1957. After graduation, Chaffee began his career as a naval aviator, both repairing and flying reconnaissance aircrafts. During the Cuban Missile Crisis, he flew 82 missions and was awarded the Air Medal. After completing a tour as an aircraft carrier pilot, Chaffee enrolled in the Air Force Institute of Technology to work on his M.S. degree in Reliability Engineering. His studies were interrupted on October 18, 1963 when he was chosen by NASA to be part of the Group 3 class of astronauts. As one of the youngest astronauts selected by NASA, Apollo 1 was to be Chaffee's first flight into space. During the flight rehearsal, he was seated in the right-most seat and charged with maintaining communications with ground controllers in the event of an emergency. Roger Chaffee was survived by his wife Martha and children, Sheryl Lyn and Stephen.

5.1.3   *"GO FEVER!"*

To meet President Kennedy's challenge of landing a man on the Moon before 1970, NASA had to develop an unprecedented amount of flight hardware, training protocols, and mission procedures in just a few short years. In this high-stakes, high-risk atmosphere, the U.S. space industry developed what is commonly referred to today as, *"Go Fever!"*— a group-think phenomenon in which people push themselves, despite great danger, to meet a previously-chosen goal. Unfortunately, *"Go Fever!"* was causing concern for the primary and backup crews of Apollo 1. For example, during a spacecraft review meeting held on August 19, 1966, the astronauts expressed worry about having so much flammable Velcro inside the cabin [5]. Despite these concerns, engineers kept the flammable material in the capsule to facilitate the securing of tools and equipment. When the Apollo 1 CM and Service Module were delivered to the Kennedy Space Center a week later, several hundred engineering changes still had to be made. Alarmed by the slow progress of these changes, the crew gave a picture of themselves posed with a model of the spacecraft to the Apollo Spacecraft Program Office Manager, Joe Shea. In the photograph, the crew's heads are bowed with eyes closed and hands clasped in prayer. The inscription reads: *"It isn't that we don't trust you, Joe, but this time we've decided to go over your head"* [5]. On December 30, the backup crew, led by Command Pilot, Wally Schirra, completed a successful altitude chamber test of the craft. After the test, a worried Schirra told Grissom: *"There's nothing wrong with this ship that I can point to, but it just makes me uncomfortable. Something about it just doesn't ring right."* Schirra also cautioned Grissom to get out of the ship at the first sign of trouble [6]. On January 22, 1967, Grissom became so frustrated with the inability of engineers to keep the training simulator in synch with the actual spacecraft, he took a lemon off a tree in his backyard and hung the lemon on the simulator [7].

5.1.4   *"WE'VE GOT A FIRE IN THE COCKPIT"*

Cape Kennedy Air Force Station Launch Complex 34A was the site for the *"Plugs Out Integrated Test"* of the AS-204 spacecraft on January 27, 1967. The *"plugs out"* moniker describes a test of the vehicle to see how it performs under internal power, with no umbilicals supplying off-board power to the ship. The rehearsal was dubbed *"non–hazardous"* since no pyrotechnic systems were armed nor was the rocket fueled. At 1:00 pm, the crew climbed into the capsule and were strapped into their seats. Grissom immediately reported a foul odor of *"sour buttermilk"* circulating through his suit. The simulated countdown was suspended at 1:20 pm and resumed at 2:42 pm when engineers could not identify a cause of the odor. At this point, the complicated, three-layered hatch was closed. The air in the cockpit was then replaced by pure oxygen, pressurized to 16.7 psi, to drive out any air that entered the cockpit as the crew boarded as well as to seal the *"plug door."* Such a door seals itself by taking advantage of a pressure difference established across its two sides. As the cabin is pressurized, a wedge-shaped door is forced into a socket, forming a seal that prevents it from being opened until the cabin pressure can be released. Most commercial aircraft use a plug door design. The decision to use a pure oxygen environment, over a duel-gas nitrogen-oxygen system, makes sense for several reasons: it is simpler to design, weighs less, and eliminates the possibility of decompression sickness, (i.e., *"the bends"*). Once the capsule was in space during an actual flight, the pure oxygen atmosphere would have been lowered to 5 psi to reduce the risk of fire while still sealing the hatch against the almost zero pressure of space.

Problems plagued Apollo 1 all afternoon; most of them involving the communications system. The countdown was suspended again at 5:40 pm while engineers tried to debug the problems. At 6:30 pm, the countdown remained on hold. Grissom's microphone was stuck on (recording much of the audio used to determine what happened that fateful day) and controllers heard him question: *"How are we going to get to the Moon if we can't talk between two or three buildings?"* In the midst of all of this *"Go Fever,"* something was about to *"go"* terribly wrong. At exactly 6:30:55 pm, engineers detected a power surge that accompanied an electrical short, probably sparked by a chafed wire, somewhere in the lower-left side of the CM near the environmental control unit below Grissom's seat. The Apollo 1 fire now had its ignition source. At 6:31:04 pm, a crew member (an audio analysis is inconclusive as to the identity) shouted: *"Hey!"* or *"Fire!"* At 6:31:06 pm, Chaffee reported: *"We've got a fire in the cockpit."* At 6:31:13 pm, a badly garbled voice (believed to be White's) shouted: *"We've got a bad fire … Let's get out … We're burning up,"* followed by a prolonged scream of pain [8].

White struggled to open the main hatch while Grissom, blocked by a wall of flames, failed to reach the emergency cabin pressure vent valve. The accident Grissom had on his Mercury mission had come back to haunt the crew of Apollo 1. To prevent the hatch from accidentally opening upon splashdown, the hatch had been redesigned with no explosive bolts to blow it open. As the fire burned, pressure inside the cabin increased, sealing the plug door tighter and tighter. At 6:31:19 pm, the interior pressure reached 29 psi and burst the inner wall of the CM, allowing ambient air into the cabin. A secondary fire broke out as flammable materials inside the cabin, (i.e., polyethylene tubing, Velcro netting, nylon suits, etc.) burned (see Figure 5.3). Smoke and lethal gases from the fire asphyxiated the astronauts. All transmissions of voice and data from the spacecraft terminated by 6:31:22, three seconds after the CM's inner wall burst.

By roughly 6:36 pm, ground controllers opened the hatch—only 5 minutes had elapsed since the first report of a fire! Emergency personnel found Grissom's seat in the 170° position, meaning it was essentially flat. He had removed his restraints, unlocked his foot restraints, and was found on the floor of the cockpit, helmet visor closed. The Apollo 204 Review Board later found that had Grissom been able to open the cabin valve, its venting capacity was insufficient to prevent the rapid buildup of pressure due to the fire—its venting would have delayed the CM's rupture by less than a second. In other words, had Grissom managed to open the cabin pressure relieve valve, his actions would not have prevented the secondary fire that asphyxiated the astronauts [8]. White's seat was in the 96° position, with the back portion horizontal and lower portion raised. Emergency procedures called for him to leave his restraints in place and attend to the hatch. His buckles were not opened, but the restraints had been disintegrated by fire. Ground controllers watching the CM on television screens testified they saw White reaching for the inner hatch handle—he had tried in vain to open the hatch. He was found lying sideways below the hatch, helmet visor closed. Chaffee remained dutifully in his seat as the fire

Figure 5.3: The badly charred interior of the AS-204 Command Module in the aftermath of the fire (Courtesy of NASA).

swept left-to-right across the cabin and attempted to stay in contact with ground controllers. His seat was in the 264° position with the back portion horizontal but the lower portion dropped to the floor. All of Chaffee's restraints were disconnected [8]. Being farthest from the ignition source, his remains were burned the least. Seven and a half hours after the fire, the bodies were removed. Their removal took over 90 minutes as the fire's heat had melted the nylon of their space suits and life-support hoses, thus fusing the bodies to the interior of the cockpit. Autopsies confirmed that all three crewmen died from carbon monoxide poisoning, resulting in cerebral hypoxia and cardiac arrest. Burns suffered by the crew were not believed to have contributed to their deaths as they occurred postmortem.

Grissom and Chaffee were buried at Arlington National Cemetery while White was buried at West Point Cemetery. After the successful launch of Apollo 7, Launch Complex 34 A was dismantled. The complex's remaining concrete pedestal now bears two plaques commemorating the crew of Apollo 1. Over the past half century, other tributes and memorials have been established to commemorate the crew. Craters on the Moon were named Chaffee, Grissom, and White. The astronauts have been inducted posthumously into the U.S. Astronaut Hall of Fame and the International Space Hall of Fame. President Carter awarded Grissom the Congressional Space Medal of Honor posthumously in 1978; President Clinton awarded White and Chaffee the same medal posthumously in 1997. Today, visitors to the Kennedy Space Center can see the names of the crew adorning the Space Mirror Memorial while visitors to Purdue University can visit two engineering buildings, Grissom and Chaffee Halls. Finally, the starship featured in the 1984 movie, *"Star Trek III: The Search for Spock,"* was the *"U.S.S. Grissom."*

After the fire, the Apollo 204 Review Board cited vulnerable wiring and plumbing as the likely source of the fire, although the specific source of the spark was never identified. Engineers subsequently implemented major modifications in the design, materials, and procedures of the Apollo program. Among these modifications were the development of improved flammability tests for materials used in crewareas of manned spacecraft and improved acceptance criteria for the flammability, odor, and toxicity of materials used in environments that support combustion [9].

5.1.5   STATEMENT OF THE PROBLEM

In our prior work, we investigated how fabrics burn *vertically* after the source of ignition is removed. The activity was modeled after ASTM D 6413-99—*The Standard Test Method for Flame Resistance of Textiles (Vertical Test)*, which has been adopted as an accepted Federal Test Standard [2, 10]. We now add a *horizontal* flame test to our repertoire of laboratory activities. This test is modeled after EPA Method 1030—*Ignitability of Solids* [11]. Although the method is used primarily to test pastes, granular materials, and powdery substances, it works on any solid material that can be cut into strips. Since the procedures for the vertical flame test were already described in our previous publication, we now describe only the procedures for the horizontal flame test, yet report results for both orientations [2]. Also, since English units are used in the ASTM protocols, we also use English units in our analysis. The equipment and materials needed to run this activity cost under $30 and were readily obtained from local hardware and fabric stores. We piloted our activity on a cohort of undergraduate students.

5.1.6   THE APOLLO CABIN

The Apollo CM was a conically-shaped capsule that was 11 feet, 1.5 inches high and 12 feet, 6.5 inches in diameter and had approximately 210 ft [3] of habitable space [12]. The Apollo 204 Review Board's reconstruction of events estimates the fire to have progressed as follows: first, the electrical short occurring near the lower-left corner of the CM probably ignited the polyethylene tubing that covered the wires running throughout the capsule. Several large patches of adhesive Velcro hooks that were attached to the wall panels of the CM, leg-rests, and seats were the next materials to ignite. Finally, when the fire reached the astronauts, the nylon outer covering of their space suits melted. Several other combustible materials were also factors in the fire; however, polyethylene tubing and Velcro (both hook and loop sides) formed the basis for our laboratory-based exercise.

5.1.7 SAMPLE PREPARATION

We purchased a 24 inches × 6 inches × 5/16 inch ceramic plate at a local hardware store. Using a permanent marker, we drew two lines, 4.0 inches apart, centered on the plate. The burn rate of samples would be measured between these two lines. We made no attempt to purchase fabrics or tubing with the exact specifications of those aboard Apollo 1. Instead, we chose to test the readily-available, modern-day versions of these materials. We are aware that the materials manufactured today are different from those aboard Apollo 1, especially since flame-retardant technologies have dramatically evolved over the last fifty years. With this in mind, we next purchased polyethylene wire covering that we rolled flat and swatches of Velcro hooks and loops that we ironed flat. These three materials were then distributed to students who trimmed them into five 12 inch × 3 inch strips and weighed each strip (the weight of each sample is only needed for the vertical burn test). Each strip was placed into a frame of sheet metal that secured the strip on its two long sides, leaving its two short edges exposed. The strips were clamped to the frame at four locations with simple binder clips. Using this sample preparation technique allows students to test samples in either the horizontal or vertical orientations.

5.1.8 TESTING AND RESULTS

We placed the ceramic plate on the floor of a safety hood. We then set the metal frame 1 inch over the center of the plate by balancing the corners of the frame on stacks of 2 inch × 2 inch ceramic tiles (Figure 5.4a). We oriented the long-axis of the frame perpendicular to the airflow of the safety hood, which was held at 0.7-to-1 meter/sec, and took precautions to minimize drafts in the hood. To enhance the filming of flames, we dimmed the lights and placed a black poster inside the hood to serve as a background. We made no attempt to control ambient temperature nor to create an oxygen-rich pressured atmosphere. Each of the three samples was tested five times and the average was reported per sample. In agreement with the literature, our burn rates were repeatable to within 10% [11]. A Bunsen burner, with 10-mm inside diameter barrel, was used to create a 1.5-inch high, 99%-pure methane, flame. The burner could be swiveled so that the tip of the flame was brought to the exposed edge of the sample. The flame was applied for 12 ± 0.5 seconds (flame-to-strip), as measured by a stopwatch. Note that EPA Method 1030 (horizontal flame test) recommends that the flame be applied for two minutes while ASTM D 6413-99 (vertical flame test) recommends that the flame be applied for only 12 seconds. In order to present students with a uniform testing procedure, we adopted the 12-second application for both types of test. Also, beware that many synthetic materials are prone to *"sputtering"*—a phenomenon whereby burning particles of fabric are sporadically ejected several inches from the propagating flame-front. Although fascinating to observe, sputtering can be especially problematic during the vertical flame test so be sure that the students completely close the window to the safety hood once they remove the flame from the sample. Students filmed each trial using cell phone cameras in *"slow motion"* mode. Once the flame was removed, students continued to film the strip until any visual flame or glow self-extinguished (Figures 5.4b, 5.4c,

Figure 5.4: Our apparatus for horizontal flame testing.

Table 5.1: Horizontal and vertical flame test results

Sample	Weight per Area ($\times 10^{-3}$)	Dripping or Melting	Sputtering	Horizontal Orientation			Vertical Orientation				
				Burn Time [over 4" path]	Burn Rate	Burn Time*	Afterflame †	Afterglow †	Char Length †	Burn Rate	Burn Time *
	lbs/in²			sec	in./sec	min	sec	sec	in.	in./sec	min
Velcro (Hook)	1.13	Yes	Yes	66	0.06	41.4	89	120	12	0.14	16.5
Velcro (Loop)	1.10	Yes	Yes	79	0.05	49.5	78	139	12	0.15	14.5
Polyethylene	1.31	Yes	Yes	140	0.03	87.8	133	190	12	0.09	24.7
*Burn times were extrapolated to appropriate dimensions of the Apollo CM. †These terms were described in our prior work.[2]											

and 5.4d). These videos are used to determine the duration of time, to 0.1-second resolution, that the samples needed to burn across the 4-inch distance marked by the permanent lines on the ceramic plate. Any signs of melting, dripping, or sputtering were noted.

As students conduct their trials, the propagation of flames, accompanied by the sputtering, dripping, and melting of fabrics and tubing, is striking to see. Even in the absence of a pressurized oxygen environment, students dramatically observe the highly dangerous conditions that must have consumed the interior of the Apollo 1 cockpit! Table 5.1 shows data from a typical run of our activity. Data from the horizontal flame tests indicate that without the pressurized oxygen environment, a fire consuming pure polyethylene or Velcro as its fuel would need 40–90 minutes to travel a distance equal to the diameter of the CM. In the case of the vertical flame tests, such a fire would need 15–25 minutes to propagate a distance equal to the height of the CM. Thus, regardless of orientation, the samples do not burn at fast enough rates to consume a ship the size of an Apollo CM in the timeframe observed during the Apollo 1 fire, (i.e., half a minute). The results support the notion that the pressurized oxygen environment dramatically changed the flammability of the interior fabrics. In fact, literature indicates that polyethylene and Velcro burn over twice as fast in oxygen at a pressure of 16.5 psi than at 5 psi. Therefore, before the inner wall of the CM ruptured, the primary fuels for the fire burned more than twice as fast as they did under the conditions for which they were evaluated [13]. Even more incredibly, transcripts from a Senate hearing indicate that the polyethylene and Velcro inside Apollo 1 may have even burned at a rate of 2.5 in/sec, which would have indeed consumed a ship the size of an Apollo CM in under a minute [14]. In short, the pressurized oxygen atmosphere made almost every material inside the cabin, even materials not normally considered highly flammable, prone to burst into flames when given a spark. Finally, we discuss with students that three phenomena would have increased the spread rate of the fire in the capsule over that measured in small scale propagation studies similar to our laboratory activities: convection currents generated by the fire itself; dripping of burning materials resulting in a liquefied molten stream that spreads the fire; and sputtering of burning materials that projects burning particles several inches from their origin.

5.1.9   *"THE KRANZ DICTUM"*

On the Monday morning following the Apollo 1 fire, flight director Eugene *"Gene"* Kranz called a meeting of his branch and flight control teams. Kranz made the following address to his teams—his address has been since referred to as, *"The Kranz Dictum"* and is considered his legacy to NASA:

Spaceflight will never tolerate carelessness, incapacity, and neglect. Somewhere, somehow, we screwed up. It could have been in design, build, or test. Whatever it was, we should have caught it. We were too gung ho about the schedule and we locked out all of the problems we saw each day in our work. Every element of the program was in trouble and so were we. The simulators were not working, Mission Control was behind in virtually every area, and the flight and test procedures changed daily. Nothing we did had any shelf life. Not one of us stood up and said, "Dammit, stop!" I don't know what Thompson's committee will find as the cause, but I know what I find. We are the cause! We were not ready! We did not do our job. We were rolling the dice, hoping that things would come together by launch day, when in our hearts we knew it would take a miracle. We were pushing the schedule and betting that the Cape would slip before we did.

From this day forward, Flight Control will be known by two words: "Tough" and "Competent." Tough means we are forever accountable for what we do or what we fail to do. We will never again compromise our responsibilities. Every time we walk into Mission Control we will know what we stand for. Competent means we will never take anything for granted. We will never be found short in our knowledge and in our skills. Mission Control will be perfect. When you leave this meeting today you will go to your office and the first thing you will do there is to write "Tough and Competent" on your blackboards. It will never be erased. Each day when you enter the room these words will remind you of the price paid by Grissom, White, and Chaffee. These words are the price of admission to the ranks of Mission Control.

5.2   CONCLUSIONS

A compelling case can be made for bringing the Apollo 1 fire into the introductory physics classroom. First, the physics of flammability can be treated appropriately at the introductory level since only careful measurements of time, distance, and weight are needed. Second, the resulting analysis provides instructors with several *"teachable moments"*—the case study: brings real-world applications of physics to the classroom; shows how standardized testing protocols are used; demonstrates how a cascade of unlikely events can result in an unpredictable catastrophe; and showcases how historical events are open to re-visitation and scrutiny. Additionally, the Apollo 1 fire truly demonstrates the "Normalization of Deviance," a phenomenon where low performance standards become acceptable because they "get the job done." Each time the low performance standards work, with no adverse effects, they become easier to accept and use again. In general, only a catastrophe or system-wide failure causes the standards to be re-examined.

In an interview conducted by Gus Grissom a few weeks before his death provides perhaps the best reason for sharing this case study with our students. During the interview, Grissom was asked about the dangers of spaceflight. He replied: *"If we die, we want people to accept it. We're in a risky business, and we hope that if anything happens to us it will not delay the program. The conquest of space is worth the risk of life"* [15]. Thus, bringing the Apollo 1 fire to the introductory physics classroom uniquely raises the historical awareness of our students by vividly portraying the heroic efforts of those individuals involved in the early days of the exploration of space.

5.3   REFERENCES

[1] G. A. DiLisi and S. McLean, The Apollo 1 fire: A case study in the flammability of fabrics, with supplemental material for on-line appendix, *Phys. Teach.*, 57:236–239, April 2019. 72

[2] G. A. DiLisi, The *Hindenburg* disaster: Combining physics and history in the laboratory, *Phys. Teach.*, 55:268–273, May 2017. DOI: 10.1119/1.4981031. 72, 78

[3] M. Cassutt and D. K. Slayton, *Deke! U.S. Manned Space: From Mercury to the Shuttle*, St. Martin's Press, New York, 1994. 74, 78

[4] NASA content used in a factual manner that does not imply endorsement may be used without needing explicit permission, December 18, 2017. https://www.nasa.gov/multimedia/guidelines/index.html

[5] C. Murray and C. B. Cox, *Apollo: Race to the Moon*, Simon and Schuster, New York, 1989. 75

[6] J. Kluger and J. Lovell, *Apollo 13*, Houghton Mifflin Company, Boston, MA, 2000. 75

[7] M. C. White, Detailed biographies of Apollo I crew—Gus Grissom, *NASA History Program Office*, NASA. 75

[8] NASA, Report of Apollo 204 review board, NASA, April 5, 1967. https://history.nasa.gov/Apollo204/summary.pdf on November 7, 2017. 76, 77

[9] NASA, *Flammability, Odor, and Toxicity Requirements and Test Procedures for Materials in Gaseous Oxygen Environments. MSFC-SPEC-101*, NASA George C. Marshall Flight Center, Huntsville, AL, August 1968. 78

[10] American Society for Testing and Materials, *ASTM D:6413–99, Standard Test Method for Flame Resistance of Textiles (Vertical Test)*, Annual Book of ASTM Standards, ASTM, West Conshohocken, PA, 2016. 78

[11] EPA, Method 1030, revision 1, July 2014, final update V to the 3rd ed., of the test methods for evaluating solid waste, *Physical/Chemical Methods*, EPA Publication SW-846, EPA, Washington, DC, 2015. 78, 79

[12] NASA, *Apollo Operations Handbook Block II Spacecraft. SM2A-03-Block II-(1)*, NASA Apollo Spacecraft Project Office, Houston, TX, April 15, 1969. https://history.nasa.gov/alsj/SM2A-03-BK-II-%281%29.pdf on December 17, 2017. 78

[13] NASA, *The Apollo Spacecraft—A Chronology. Vol. IV, Part:1*, Mar/Apr 1967. https://www.hq.nasa.gov/office/pao/History/SP-4009/v4p1h.htm on December 17, 2017. 82

[14] United States Senate, *Apollo Accident, Hearing before the Committee on Aeronautical and Space Sciences*, U.S. Government Printing Office, Washington, DC, February 7, 1967. https://books.google.com/books?id=toeN-glF_nsC&dq=velcro+burning+rate&source=gbs_navlinks_s on December 17, 2017. 82

[15] C. G. Brooks, J. M. Grimwood, and L. S. Swenson, Jr., *Chariots for Apollo: The NASA History of Manned Lunar Spacecraft to 1969*, Dover Publications, Mineola, NY, 2009. 83

CHAPTER 6

Demonstrating "The Perfect Storm Scenario"

Case studies demonstrate the "perfect storm scenario"—how a progression of events often results in an unlikely or unforeseen outcome

Our article on the sinking of the *S. S. Edmund Fitzgerald* provides the basis of this case study. On November 10, 1975, the Great Lakes bulk cargo freighter *S. S. Edmund Fitzgerald* suddenly and mysteriously sank during a winter storm on Lake Superior (see Figure 6.1) [1]. All 29 men onboard perished. Students see that an unlikely cascade of conditions, (i.e., the *"perfect storm scenario"*) were met; thus, placing the ship at the wrong place at the wrong time. We present the historical debate as: *"Why did the S. S. Edmund Fitzgerald sink in Lake Superior on November 10, 1975?"*

6.1 REMEMBERING THE S. S. EDMUND FITZGERALD: A CASE STUDY IN ROGUE WAVES

Gregory A. DiLisi, *John Carroll University, University Heights, Ohio*
Richard A. Rarick, *Cleveland State University, Cleveland, Ohio*

6.1.1 INTRODUCTION

November 10th marks the anniversary of the sinking of the *S. S. Edmund Fitzgerald* [2], a Great Lakes bulk cargo-freighter that suddenly and mysteriously sank during a severe winter storm on Lake Superior. The entire crew of 29 men perished. The ship sank so quickly that its captain never even issued a distress call. A year after the sinking, Canadian folksinger Gordon Lightfoot wrote and recorded the ballad, "The Wreck of the Edmund Fitzgerald." The song became an international hit single that sparked worldwide interest in the ship and its crew. Since then, the sinking of the *S. S. Edmund Fitzgerald* has become the most well-known and controversial shipping disaster on the Great Lakes.

The purpose of this chapter is to commemorate the anniversary of this tragic event by bringing it to the attention of a new generation of students, namely, those enrolled in our introductory physics classes. We focus our analysis on the formation of rogue waves as a possible explanation of why the massive freighter sank. Our treatment is a compelling fit for introductory physics classrooms.

Figure 6.1: The S. S. Edmund Fitzgerald [1].

Since most of our students were not yet born when the ship sank, we first establish a historical context for them by providing detailed information about the ship, its final voyage, its underwater wreckage site, and the four prevailing theories of why it sank. (Lyrics from Lightfoot's ballad headline each of these sections to more artistically portray historical events.) Next, we focus on one of these theories and bring the topic of "rogue waves" to the introductory physics classroom. Finally, we use the "wave-focusing" theory of rogue wave formation, coupled with the principle of super-position, to produce a simple simulation of the conditions that might have resulted in the giant freighter's sudden sinking.

6.1.2   THE MIGHTY FITZ

The ship was the pride of the American side
coming back from some mill in Wisconsin.
As the big freighters go, it was bigger than most
with a crew and good captain well seasoned [3].

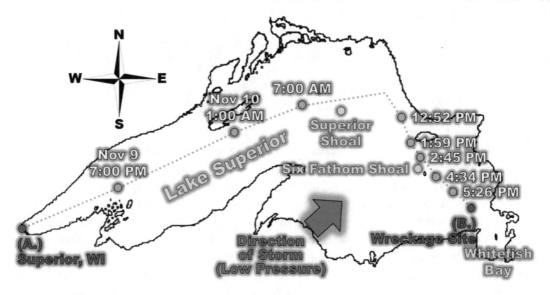

Figure 6.2: Approximate path of the final voyage of the *S. S. Edmund Fitzgerald*.

The *S. S. Edmund Fitzgerald* was the flagship Laker-type freighter of the Columbia Transportation Fleet and was named after the president and chairman of the board of The Northwestern Mutual Life Insurance Company, which had paid to have it built because of the company's heavy investments in the mining and shipping industries of the Great Lakes region. The freighter was constructed to be a massive floating cargo-hold. A multi-story pilot-house at the bow of the ship was connected to the stern's engine-room via three separate cargo-compartments, making the overall cargo-hold 500-ft in length. Cargo was accessible via 21 top-side hatches. At the time of its launch in 1958, it was the longest freighter on the Great Lakes and quickly became known as a workhorse for its seasonal record-setting hauls of taconite iron ore.

By 1975, the Fitzgerald had made an estimated 750 round-trips across the Great Lakes, earning it the nicknames: "Queen of the Lakes," "The Mighty Fitz," and "The Big Fitz."

On what would be its final voyage, *"The Fitz,"* as it was called by its crew, left Superior, Wisconsin (see Figure 6.2—Point (A)) at 2:15 pm on November 9, 1975 under the command of Captain Ernest M. McSorley, who had been captain of the Fitz for the previous three shipping seasons. McSorley told his wife that he planned to retire at the end of that shipping season. The ship was carrying 26,000 tons of taconite ore bound for a steel mill near Detroit, Michigan. Though a severe storm was predicted to pass just south of Lake Superior during the morning hours of November 10, the Fitz actually began its voyage under calm seas and clear skies. By 5:00 pm, the Fitz began following another freighter, the *S. S. Arthur M. Anderson,* commanded by Captain Jesse B. Cooper. The two freighters would travel within a few miles of one another and stay in constant radio (and intermittent visual) contact for the next day.

6.1.3   THE FINAL VOYAGE

The captain wired in he had water comin' in
and the good ship and crew was in peril.
And later that night when 'is lights went outta sight
came the wreck of the Edmund Fitzgerald [3].

By 1:00 am on November 10, the winter storm had changed course and was sweeping to the northeast across Lake Superior. McSorley and Cooper responded by also changing course so that the two freighters could gain cover by hugging the Canadian coastline of Lake Superior. At first, the course-change provided the ships with some much needed protection, but by 1:40 pm, the winds shifted from northeasterly to northwesterly (wind direction refers to the direction from which the wind is coming) and massive waves began hitting the ships. The faster Fitz pulled about 16 miles ahead of the Anderson and as freezing rain and snow pelted the two ships, the Anderson soon lost visual contact with the Fitz. At 3:30 pm, now heading southeast and hoping to reach the relative safety of Whitefish Bay, McSorley reported that the Fitz was taking on water and had developed a severe starboard-list. In response, McSorley slowed his ship so that the Anderson could pull to within 10 miles. At 4:10 pm, McSorley reported that he had lost use of both of the ship's radars and had activated two of the pumps to vent water that was accumulating in the ballast tunnels that ran along the length of the ship. The Fitz was now sailing blind through a hurricane-force storm—waves as high as 35 feet were reported and sustained winds were recorded between 58–67 mph, with gusts as high as 86 mph. The severity of the storm can be judged by one of McSorley's last radio communications, made at approximately 5:30 pm. McSorley, a *40-year veteran* mariner of the Great Lakes, reported: "I have a bad list. I have lost both radars, and am taking on heavy seas over the deck in one of the worst seas I've ever been in." Sadly, at 7:10 pm, McSorley optimistically radioed to Cooper that, "We are holding our own," but these would be the last words transmitted by McSorley. Minutes later, the Fitz mysteriously disappeared from Cooper's radar and all radio contact with the Fitz was lost. No distress call had been sent by McSorley.

6.1.4   THE WRECKAGE SITE

Does anyone know where the love of God goes
when the waves turn the minutes to hours?
The searchers all say they'd have made Whitefish Bay
if they'd put fifteen more miles behind 'er [3].

On November 14, 1975, Navy aircraft, equipped with magnetic sensors, found the Fitz, with its hull broken in two, under 530 feet of Canadian waters at the bottom of Lake Superior, approximately 17 miles west of the entrance to Whitefish Bay (see Figure 6.2—Point (B)). Some debris, including shattered lifeboats and rafts, was eventually found washed ashore but

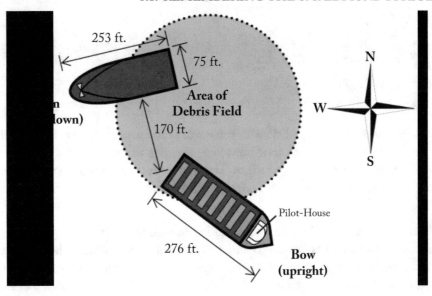

Figure 6.3: Approximate location of the wreckage and debris field at the bottom of Lake Superior.

no bodies were ever recovered. The Fitz remains the largest ship to have sunk on the Great Lakes. Details of the wreckage site prove to be as much of a mystery as the circumstances of the ship's disappearance (see Figure 6.3): A 276-ft × 75-ft section of the bow lies upright facing southeast while 170 feet to the north lies a 253-ft × 75-ft section of the of the stern facing northeast, but turned completely upside down. The missing 200-ft of the mid-ship lies in an approximately circular debris field, 200-ft in radius and primary filled with taconite ore pellets, twisted metal, and hatch-coverings. Regardless of what caused the ship to sink, it obviously suffered some type of catastrophic structural failure and split in two before hitting, or as it hit, Superior's bottom. In addition, the debris field indicates that once the ship split in two, taconite ore pellets from the stern-half spilled out of its cargo-hold and hit the bottom first. The two halves of the ship then landed atop the pellets. All 29 members of the crew are presumed to rest within the wreckage. The Ontario government now officially recognizes the wreckage as a "watery grave," thus preventing any dives on, or surveys of, the site without a license. At the request of the crew's families, the ship's 200-lb bronze bell was recovered on July 4, 1995 and is now on display at the Great Lakes Shipwreck Museum, located at the Whitefish Point Light Station near Paradise, Michigan.

6.1.5   THEORIES

They might have split up or they might have capsized;
they may have broke deep and took water.
And all that remains is the faces and the names
of the wives and the sons and the daughters [3].

Despite multiple underwater expeditions to the wreck, the cause of the sinking remains a mystery. For years, scientists put forth four prevailing theories to explain how the Fitz ended up more than 500 feet below the water's surface. Today, a re-examination of the evidence, coupled with testimony from the crew of the *S. S. Arthur M. Anderson*, now appears to discount three of these theories.

- **Structural failure:** The Fitz had recently undergone structural modifications that allowed it to carry heavier cargo and travel lower in the water. Had these modifications weakened the ship's hull to the point where it fractured under the stresses of the November storm? According to this theory, the ship literally broke in half *while on the surface* of the water. However, finding the two sections only 170 feet apart indicates that the Fitz probably started its descent in one piece and broke apart either when it hit the bottom or well after it started its descent.

- **Shoaling:** The ship's path took it dangerously close to the Six Fathom Shoal and Superior Shoal, the highest points of which lie only 20–30 feet below the water's surface (see Figure 6.2). Perhaps the Fitz bottomed-out on one of these shoals, causing water to leak slowly into the cargo-holds or ballast tanks until the ship's buoyancy was catastrophically compromised. The Lake Carriers Association (LCA) concluded that the ship had indeed bottomed-out at the Six Fathom Shoal and most likely sank from the ensuing damage done to the ballast tanks [4]. However, examinations of both shoals and the overturned stern section of the ship, show no telltale signs of damage to the rudder, propeller, or shoals.

- **Opened hatch-coverings:** Once the ship set sail, the Fitz's deckhands were supposed to seal the top-side hatches by bolting watertight covers over them. The ship may have sunk from loss of buoyancy resulting from water breaching these hatch-covers and accumulating in the cargo-hold. Subsequent photographic evidence from the wreckage site shows *no damage* to the bolts that were supposed to secure these hatch-covers, suggesting that the bolts were never properly secured. If the bolts were in fact secured, they would have been severely mangled with the rest of the mid-ship, as the ship's hull tore in two. The official U.S. Coast Guard (USCG) and National Transportation Safety Board (NTSB) casualty reports eventually agreed that the ship did indeed sink from water pouring into its hatch-covers [5, 6]. Because this theory blames the ship's demise on the deckhands whose job it was to secure the hatch-covers, these findings have been disputed by the surviving families as well as by the LCA. A strong refutation of this theory lies in several of Captain McSorley's communications in which he stated that he and the crew were battling the severe starboard-list. If the cargo-hatches were improperly secured, water would have been uniformly distributed in the cargo-hold, not just to the starboard side. Also, McSorley indicated that he and the crew had activated two of the ballast pumps, thus suggesting that they had already determined that the list was caused by a breach of the ballast tanks, not the cargo-hold.

6.1.6 ROGUE WAVES

The fourth and most likely explanation of events is that the Fitz, already listing to starboard, simply rolled over as it fell victim to massive waves. A relatively new area of physics research presents an even more captivating version of this explanation. "Rogue waves" (also called "freak" or "monster" waves) may have hit the ship from the stern and capsized it. In fact, maritime folklore tells of, "The Three Sisters of Lake Superior"—a set of three massive waves that pummel ships in rapid succession. Supporting this theory is Captain Cooper who testified that the *S. S. Arthur M. Anderson* was hit by two massive waves at approximately 7:00 pm. Cooper stated that: "The second large sea put green water on our bridge deck! This is 35 feet above the waterline." (To put the overall height of these waves in perspective, Cooper's mention of "green water" refers to the middle part of the wave.) Had these waves, and possibly a third one, continued to the southeast and caught up with the Fitz 10 minutes later and overwhelmed the already struggling ship?

Unfortunately, rogue waves are not the stuff of lore but are in fact, very real. Though several reported sightings of rogue waves exist as far back as the mid-1800s, the first rogue wave to be measured by a scientific instrument did not occur until New Year's Day, 1995 at the Draupner gas platform in the Norwegian North Sea. On that day, a single, giant wave measuring almost 85 feet from trough to crest (now commonly referred to as, "The New Year's Wave"), was recorded by instruments aboard the platform [7]. This event brought rogue waves out of maritime myth and into scientific reality! Though often called "monster waves," rogue waves are not necessarily the highest nor most destructive of waves. Instead, a rogue wave is an abnormally large wave relative to a statistical average of nearby waves. Physicists define a rogue wave in terms of the Significant Wave Height (SWH) of the current sea state, where the SWH is defined as the average of the largest one-third of waves in a given sampling of waves. A "rogue wave" is defined as one whose height is more than 2.2 times as high as the SWH [8].

The study of rogue waves is a robust area of current physics research. With applications that extend beyond hydrodynamics to fields of study such as optical supercontinuum generation, nano-devices, energy management, advanced imaging, and environmental safety, numerous studies are now underway attempting to better understand and replicate the necessary conditions for the formation of these waves. However, because rogue waves are relatively new to the scientific community, the conditions that produce them are simply not well understood. At first glance, one might think that strong winds are responsible for generating rogue waves; however, research shows that wind alone is simply not powerful enough to generate them. Thus, wind is more commonly viewed as a contributing factor to the formation of rogue waves than the actual cause. Instead, researchers have posited several alternative theories to explain the formation of rogue waves. These theories have been simulated computationally or verified experimentally in wave tanks and optical fibers, yet each theory is plagued by its inability to convincingly explain how its necessary conditions might actually occur in nature.

- **Wave-Focusing:** Because the velocity of water waves depends on wavelength and depth, a localized region may exist where waves shorten, thus increasing their amplitudes and decreasing their speeds. The decrease in speed allows faster waves to catch up, resulting in a coherent "focusing" of waves in this small region. Current- or temperature-gradients may also contribute to the focusing phenomenon.

- **Diffractive Focusing:** Diffraction from coastal shapes and sea-floor topology causes small waves to constructively interfere into a freak wave.

- **Nonlinear Effects:** The Nonlinear Schrödinger wave equation has been applied to predict the time evolution of water waves. The nonlinear term in the equation can lead to the formation of rogue waves under certain conditions. In short, the nonlinear mechanism allows one wave to gather energy from nearby smaller waves until it becomes of rogue-like proportions [9, 10].

- **Normal Waves:** Rogue waves are not statistical "freak waves" after all; in fact, they are part of the regular distribution of wave heights for a given sea state, just at the extreme, rarely seen limit.

6.1.7   STATEMENT OF THE PROBLEM

For several reasons, the 40th anniversary of the sinking of the *S. S. Edmund Fitzgerald* brings compelling teaching opportunities to the introductory physics classroom: first, the physics of rogue waves can be treated at the appropriate level. Only two concepts are needed: the "wave-focusing" theory of rogue waves and the principle of superposition. Since the superposition principle applies to topics from all areas of physics, rogue waves present us with a unique chance to reinforce this concept with our students. Regardless of one's particular curriculum or location in the sequence, teachers of introductory physics should find the topics discussed in this article as an appropriate fit to their classrooms. Second, the anniversary is a powerful way of raising historical awareness in our students while bringing real-world applications of physics to them. Third, the formation of rogue waves is a dynamic area of front-line physics research. By examining rogue waves in class, we emphasize the importance of staying current with recent developments in scientific research and demonstrate how this informs our teaching. Finally, the resulting analysis shows students that the answers to complex questions are not always known and sometimes remain a mystery.

Rogue waves can be cast as a real-life example of the principle of superposition in action. To start, we present "wave-focusing" to our students as a scenario in which two water waves, with different wavelengths, travel in the same direction but with different speeds. Initially, the longer wave trails the shorter one. The speed of waves on water is wavelength-dependent, according to the equation:

$$v = \sqrt{\frac{g \cdot \lambda}{2\pi} \tanh\left(2\pi \frac{d}{\lambda}\right)},$$

where d is the water depth and λ is the wavelength. Since the longer wave travels faster than the shorter one (especially in the limit of "deep water" where $d > \lambda/2$ and $v \approx \sqrt{\frac{g \cdot \lambda}{2\pi}}$), the two waves will eventually meet. As the longer wave overtakes the shorter one, the two waves will superimpose briefly to form a large-amplitude wave. When these two waves are also superimposed upon a background sea state, a "monster-type" wave may develop. Before attempting to simulate these conditions for students, one might anticipate a skeptical student asking: "Do these conditions actually occur in nature?" As mentioned above, contemporary research posits several potential causes for wave-focusing, (i.e., dispersion, current- or temperature-gradients, etc.), but we de-emphasize the actual underlying physical mechanism. To our students, we simply state that the described waveforms are necessary conditions for the formation of rogue waves, but that the causes of these conditions are still under investigation as an active area of physics research.

To demonstrate the scenario in class, we wrote a simple code in *Microsoft Excel*[TM]. Ocean waves are often simulated by sinusoidal waves, but experimental evidence suggests more realistic waveforms to be "trochoids"—more specifically "common trochoids" (for low amplitude waves) or "curtate trochoids" (for high amplitude waves). Therefore, in our simulation, the initial profile of each of the two traveling waves was a single cycle of a trochoid. The long wave was initially located behind (to the left of) the short one. The initial profile of the sea state was generated by the waveform:

$$y(x)_{sea} = y_{sea_0} \cdot Rnd \cdot \cos(k_{sea} \cdot x),$$

where Rnd is a random number between 0 and 1, called at every value of x, and y_{sea_0} is a constant that is arbitrarily chosen to set the amplitude of the initial sea state. We wrote our code so that the user could also control the initial amplitude, wavelength, location, and velocity of all waveforms. To propagate the waves in time, a scrollbar was inserted to control t, which in turn changed x according to: $x = x_0 + v \cdot t$, where x_0 is the initial position of each wave. The two waves were propagated to the right but the longer wave was set to propagate faster than the shorter one. One advantage of using a computer simulation (vs. a wave tank, say) is that no actual physical mechanism is needed to cause waves to propagate at different speeds. The sea state was propagated to the left only because most reports of rogue waves indicate that waves were traveling against the prevailing current. (This condition is not necessary for the simulation

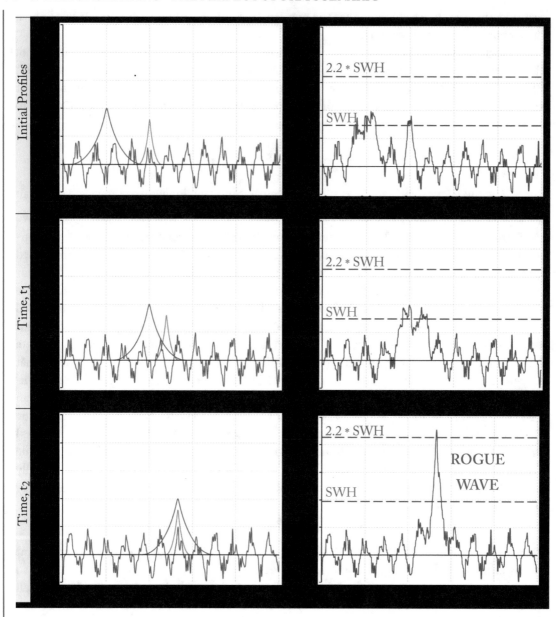

Figure 6.4: Results from our simulation. Column 1 depicts the individual waveforms and background sea state. Column 2 depicts the composite of the waves shown in Column 1. Column 2 also indicates the SWH of the composite waves and the height at which a wave becomes "rogue." At time, t_2, the longer wave overtakes the slower wave and a rogue wave forms near $x \approx 0.52$. (*Continues.*)

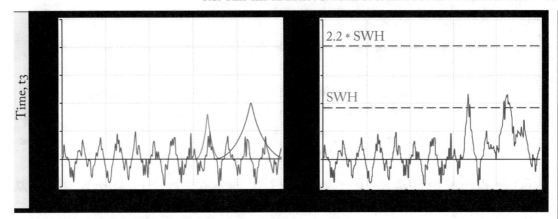

Figure 6.4: (*Continued.*) Results from our simulation. Column 1 depicts the individual wave-forms and background sea state. Column 2 depicts the composite of the waves shown in Column 1. Column 2 also indicates the SWH of the composite waves and the height at which a wave becomes "rogue." At time, t_2, the longer wave overtakes the slower wave and a rogue wave forms near $x \approx 0.52$.

to produce the desired results.) No attempt was made to account for dispersion or damping in the propagation of the waves. Other useful numerical techniques, employing a finite difference form of the wave equation, are discussed in the literature [11].

6.1.8   RESULTS

Figure 6.4 displays our results. In Column 1, we plotted the two waves (red and green curves) and background sea state (magenta curve) at different time-steps. Column 2 highlights the principle of superposition where we plotted the composite of all three waves (blue curve) as well as the SWH of the composite waves (blue dashed line) and height that these waves must reach to qualify as "rogue" (red dashed line). Row 1 displays the waveforms in their initial state. The longer waveform clearly trails the shorter one and the composite plot shows a series of regular sea waves. Row 2 displays the waveforms when the longer wave is beginning to catch up with the shorter one. At this instant, the composite plot still shows a series of regular sea waves. Row 3 displays the results at the instant the longer wave overtakes the shorter wave. At this instant (the phenomenon only occurs at a few time-steps), and in a localized region, (i.e., near $x \approx 0.52$), wave-focusing occurs and a large rogue wave forms with a height that exceeds the SWH of the composite waves by a factor greater than 2.2. Since the rogue wave forms in such a localized region, and for such a brief instant of time, students come to the conclusion, *perhaps correctly*, that the Fitz was just "at the wrong place, at the wrong time." Row 4 shows a later time when the longer wave has traveled well past the shorter one and the wave-focusing effect has disappeared. The composite plot once again shows a series of regular sea waves.

6.2   REFERENCES

[1] Photograph courtesy of Wikimedia, May 29, 2020. mons.wikimedia.org/w/index.php?curid=36483480 85, 86

[2] G. A. DiLisi and R. A. Rarick, Remembering the S.S. Edmund Fitzgerald, *Phys. Teach.*, 53:521–525, December 2015. DOI: 10.1119/1.4935760. 85

[3] G. Lightfoot, The wreck of the Edmund Fitzgerald (recorded by Lightfoot, G.), *Summertime Dream*, (LP) Toronto, ON, Reprise Records, 1976. 86, 88, 90

[4] H. Bishop, *The Night the Fitz Went Down*, Lake Superior Port Cities, Duluth, MN, 2000. 90

[5] United States Coast Guard, Marine board casualty report: *SS Edmund Fitzgerald*; sinking in Lake Superior on November 10, 1975 with loss of life, July 26, 1977. 90

[6] National Transportation Safety Board, Marine accident report: *SS Edmund Fitzgerald*; sinking in Lake Superior on November 10, 1975, May 4, 1978. 90

[7] D. A. G. Walker, P. H. Taylor, and R. Eatock Taylor, The shape of large surface waves on the open sea and the Draupner New Year wave, *Appl. Ocean Res.*, 26:(3–4):73–83, May–June, 2004. DOI: 10.1016/j.apor.2005.02.001. 91

[8] A. Chabchoub, N. P. Hoffmann, and N. Akhmediev, Rogue wave observation in a water wave tank, *Phys. Rev. Lett.*, 106:204502, May, 2011. DOI: 10.1103/physrevlett.106.204502. 91

[9] B. Guo, L. Ling, and Q. P. Liu, Nonlinear Schrödinger equation: Generalized Darboux transformation and rogue wave solutions, *Phys. Rev. E*, 85:026607, February, 2012. 92

[10] P. Dubard, P. Gaillard, C. Klein, and V. B. Matveev, On multi-rogue wave solutions of the NLS equation and position solutions of the KdV equation, *Eur. Phys. J. Spec. Top.*, 185(1):247–258, July, 2010. DOI: 10.1140/epjst/e2010-01252-9. 92

[11] N. Giordano and H. Nakanishi, *Computational Physics*, 2nd ed., pages 156–180, Prentice Hall, Upper Saddle River, NJ 2006. 95

CHAPTER 7

Developing Simulations and Testing Analogs and Proxies

Case studies showcase that scientists and engineers must often develop simulations or test analog materials in lieu of actual substances—especially if those substances are prohibitively rare, precious, expensive, or dangerous to test

In Chapter 4, we analyzed the flammability of the fabrics used to cover the German zeppelin *Hindenburg*. The flammability of the outer fabrics plays a pivotal role in the current debate surrounding the ship's destruction. However, several factors complicate a direct test of the ship's fabrics. First, obtaining a surviving piece of the ship's fabric, for the purposes of lighting in on fire, is probably out of the question. Next, the exact formula of the compounds used to dope the fabrics is not known. Last, purchasing chemicals is expensive and heavily regulated. Therefore, reproducing the *Hindenburg's* outer fabrics is simply not possible for most scientists or teachers. Instead, we chose to test three fabrics that were easy-to-make, inexpensive to buy, yet nicely approximated the outer coverings of the airship. Likewise in Chapter 5, we tested the flammability of the fabrics inside the *Apollo I* capsule: Polyethylene and Velcro (hook and loop). Knowing that the materials manufactured today are different from those aboard *Apollo I*, we chose to test the readily-available, modern-day versions of these materials.

In this chapter, we examine the conditions of the Indian Ocean Tsunami following the Sumatra—Andaman earthquake of December 26, 2004 (see Figure 7.1). As a result of the earthquake, a 1,200-km × 900-km area of the ocean floor slipped 15 vertical meters where the Indo-Australian plate subducts under the smaller overriding Burma microplate. The ensuing tsunami led to the deaths of more than 200,000 people and devastated parts of Indonesia, Sri Lanka, India, Thailand, Somalia, and Burma. Unable to re-create the exact conditions of the tsunami (We would need quite a large laboratory to re-create a tsunami, right?), we rely on two simulations to understand the event. We present the historical debate as: *"What factors contributed to the formation of the Indian Ocean Tsunami of 2004?"*

Figure 7.1: Waves from the 2004 Indian Ocean tsunami hit Ao Nang, Krabi Provine, Thailand, on December 26, 2004 [1].

7.1 MODELING THE 2004 INDIAN OCEAN TSUNAMI FOR INTRODUCTORY PHYSICS STUDENTS: A CASE STUDY IN THE SHALLOW WATER WAVE EQUATIONS

Gregory A. DiLisi, *John Carroll University, University Heights, Ohio*
Richard A. Rarick, *Cleveland State University, Cleveland, Ohio*

7.1.1 INTRODUCTION

In this article, we develop materials to address student interest in the Indian Ocean tsunami of December 2004 [2]. We discuss the physical characteristics of tsunamis and some of the specific data regarding the actual 2004 event. Finally, we create two simulations (or analogs) of the event: (1) an easy-to-make *"tsunami tank"* to create actual waves of water in a trough and (2) a somewhat simple computer-generated model. The simulations exhibit three dramatic signatures of tsunamis, namely, as a tsunami moves into shallow water, its amplitude increases, its wavelength and speed decrease, and its leading edge becomes increasingly steep as if to *"break"* or *"crash."* Using our simulations, realistic features are easy to observe in the classroom and evoke enthusiastic responses from our students.

7.1.2   STATEMENT OF THE PROBLEM

We wanted to create a *"current events"* exercise that is appropriate for an introductory physics course but that could also be used as a demonstration in a *"Physics and Society"*-type course. Given student concerns over such natural disasters as hurricanes *Katrina* and *Rita*, we decided to model the Indian Ocean tsunami that struck India, Sri Lanka, East Africa, and Southeast Asia in December 2004. The goals of our activity are as follows.

1. Discuss the physical characteristics of tsunamis and the conditions under which they are generated.

2. Present specific data regarding the 2004 event.

3. Create simulations in-class. To be realistic, the simulations must display the characteristic features of a tsunami; namely, as the water-depth decreases, the tsunami's amplitude increases while its wavelength and speed decrease.

Our activity offers several opportunities to teachers of introductory physics: it is a timely chance to expose students to fluid dynamics, a topic not covered in the standard introductory physics curriculum; it focuses on the basic physical principles of conservation laws so the concept-level is appropriate; and the context is especially engaging to students since it involves current events and has real-world applications. We stress that our primary goal is to show students how physics can be applied to a real-world event while engaging them, when appropriate, in a detailed description of water wave models.

7.1.3   CHARACTERISTICS OF A TSUNAMI

The term *"tsunami"* comes from two Japanese characters, *"tsu"* meaning *"harbor,"* and *"nami"* meaning *"wave."* Tsunamis are usually triggered by undersea earthquakes, but in general are created by any large impulsive vertical movement of the ocean floor that displaces the overlying water from equilibrium. As gravity returns the ocean to its equilibrium state, a tsunami radiates outward from the site of the displaced water.

Over most of its journey, a tsunami's physical characteristics are determined by conditions that exist in the deep ocean rather than at a shore. To understand these characteristics, consider a typical tsunami *traveling in the deep ocean*, a situation that is very different from the violent images we see on newscasts of a tsunami striking a coast. Typically, an undersea earthquake vertically shifts an area of the ocean floor by only a few feet; however, this area can be as large as a million square-kilometers, as was the case for the 2004 Indian Ocean tsunami. The huge volume of displaced ocean water relaxing to equilibrium rapidly produces a wave that is very long but not very high. In fact, in ocean depths of a few kilometers, most tsunamis have wavelengths in excess of 500 km yet amplitudes of only a few feet. For this reason, deep ocean tsunamis can easily go undetected as they are often unfelt aboard ships or unseen from the air. Only as a tsunami enters very shallow water does its amplitude increase to the dramatic heights we see on newscasts. Contrary to popular images, we find that over most of its path, a tsunami's wavelength is very large compared to the water-depth while its amplitude is very small compared to the water-depth. The horizontal scale of a tsunami dominates its vertical scale and even the deepest parts of the ocean are *"relatively shallow"* when compared to the horizontal dimensions of a tsunami. Because of these characteristics, physicists typically classify tsunamis as "Shallow Water Waves" (SWWs).

Calling tsunamis *"Shallow Water Waves"* can seem inconsistent to students; after all, our premise does sound contradictory: *"Model deep ocean tsunamis as shallow water waves."* Because tsunamis originate, and mostly travel in, *"deep"* water and are large-scale, destructive phenomena, students intuitively, but incorrectly, think that tsunamis have the characteristics of *"deep"* water waves. In general, a *SWW* is defined as any water wave that has the following physically equivalent characteristics [3]:

1. The ratio of amplitude-to-water-depth is negligible compared to the ratio of wavelength-to-water-depth.

2. The water is in *"hydrostatic equilibrium"* at all times; that is, the weight of any small volume of water is balanced by the pressure-gradient force along the vertical direction.

3. The horizontal velocity of the wave depends only on the water-depth: $v = (v_x, v_y, v_z) = (u, v, 0)$.

The physical equivalence of these three assumptions is what we believe to be the most useful to readers and the most important insight for readers to gain regarding this exercise. Consider whether a tsunami is a *"deep"* or *"shallow"* water wave … conduct a demonstration where a deep ocean tsunami is modeled as a large rolling wave generated by gently rocking a thin cookie tray of water. Using this demonstration, one can argue that the ocean behaves as a thin layer of water and a tsunami as a small-amplitude, long-wavelength *surface wave*, with little oscillation occurring over the vertical direction (Characteristic I). If the wave's amplitude-to-water-depth ratio is negligible compared to its wavelength-to-water-depth ratio, we can assume that the vertical motion, and consequently the net vertical force, is negligible. Therefore, the water is in hydrostatic equilibrium (Characteristic II). This characteristic can be difficult to reconcile with one's intuition because real waves do move vertically, contrary to what the model represents. The point here is that the horizontal scale so dominates the vertical scale, that vertical motion can be ignored. Finally, one can show mathematically that, in addition to a constant vertical pressure gradient, hydrostatic equilibrium also implies that the horizontal pressure gradients, and therefore the horizontal fluid motion, depend only on water-depth (Characteristic III). This characteristic reduces the three-dimensional wave to a simpler two-dimensional surface phenomenon. These results are derived in a number of sources in the literature. We have derived them in our upcoming computer-based simulation and leave the reader to determine the level of these materials that is appropriate for his or her use.

7.1.4   THE INDIAN OCEAN AND THE 2004 TSUNAMI

To realistically simulate the 2004 Indian Ocean tsunami, we utilized some parameters from the actual event. First, we modeled the Indian Ocean based on the bathymetric data shown in Figure 7.2 [4]. In terms of breadth, the Indian Ocean is nearly 10,000 km wide at the southern tips of Africa and Australia and covers a surface area of nearly 69 million km². The Indian Ocean floor lies roughly 7,000 m below sea-level with the deepest known point at 7,725 m, off the southern coast of Java, just south of the epicenter of the December 2004 earthquake. The figure also shows a predominant ridge along the ocean floor, known as the Mid-Indian Ocean Ridge, extending south from India and following the shape of an inverted-Y. This ridge reaches roughly 3,300 m above the ocean floor, giving the Indian Ocean an average depth of 3,900 m, with a few peaks breaking the ocean surface to form islands.

Next, let us look at the actual undersea earthquake that generated the 2004 tsunami. On December 26, 2004 at 00:58:53 UTC, an underwater earthquake, with an estimated magnitude between 9.1 and 9.3, generated the tsunami. The earthquake was the fourth largest recorded since 1900 and its epicenter was located just north of Simeulue Island, 160 km west of Sumatra (Latitude 3.316°N, Longitude 95.854°E), at a depth of 30 km below the ocean floor [6]. The U.S. Geological Survey estimates that as a result of the earthquake, a 1,200 km × 900 km area of the ocean floor vertically slipped 15 m along the interface where the dense Indo-Australian plate subducts under the smaller overriding Burma micro-plate [6]. This slip occurred discontinuously in time as the rupture proceeded northwesterly for a period of ~100 seconds, paused for another ~100 seconds, and then continued northward [6]. The overlying Indian Ocean was also vertically displaced along the entire 1,200 km-length of the rupture and waves then radiated outward. The total energy released by the earthquake is estimated to be ~2.0 × 1018 Joules. A large aftershock of magnitude 7.1 was recorded in the days that followed while numerous aftershocks, as high as 6.6, continued to shake the area for weeks after the initial rupture [6]. The resulting tsunami caused more casualties than any other in recorded history. Current estimates place victims of the tsunami at roughly 230,000–280,000 dead or missing, with an additional 5 million people displaced (see Figure 7.3).

7.1.5   SIMULATION 1 – USING A *"TSUNAMI TANK"*

7.1.5.1   Building a Tank

The highlight of our activities is simulating tsunamis with a trough of water. For this, we constructed a *"tsunami tank"*—a trough with dimensions 6 ft L × 4 in W × 10 in H. Our apparatus can be built for under $50 but did take several hours to assemble. Nowadays, these troughs are available directly for purchase from a number of vendors of scientific equipment or instructional hardware. The floor and ends of the tank are made of a U-shaped wooden frame constructed of 3/4-in thick boards held together with construction adhesive and L-brackets. The two sides are made of 1/8-in thick clear acrylic sheets custom cut by a glass shop for $20 per sheet. Grooves were cut into the wooden frame so that the acrylic sheets could be snugly inserted into the frame

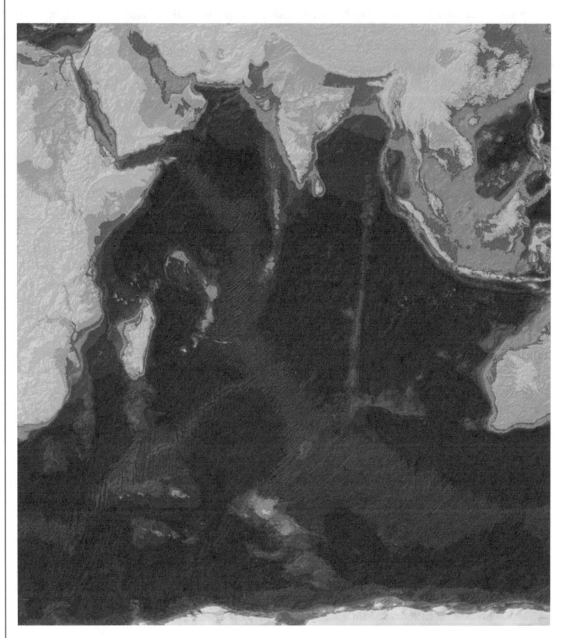

Figure 7.2: Bathymetric plot of the Indian Ocean [5].

Figure 7.3: A mosque is left standing amid the rubble in Banda Aceh. Several mosques survived and may have been saved by the open ground floor that is inherent to their design. The tsunami waves reached the middle of the second floor [7].

and glued in place with a silicone sealant. The key to successful in-class simulations is to build a tank with both deep and shallow water depths. To accomplish this, we simply tilted the tank so that the flat bottom sloped linearly from the *"deep ocean"* to a *"shallow coast."* We added blue food-coloring to the water and hung a yellow backdrop against the laboratory wall to make the waves easier to see. We found three ways to successfully simulate the earthquake and generate a tsunami-like wave: (i) we placed a thin metal plate (8-in L × 4-in W) on the floor of the deep portion of the tank, then quickly pulled the plate upward and out of the water; (ii) we held the same metal plate at the surface of the water and simply dropped it into the deep portion of the tank (this worked best); and (iii) we lifted the entire frame at the deep end of the tank and gave it an impulsive shake [8]. Two video cameras were positioned along the front of the tank and a ruler was taped to the back so we could measure amplitudes, wavelengths, and velocities. The cameras and ruler were used only to capture images for this article—they are not needed for

Figure 7.4: (a) Construction of the *"tsunami tank."* A trough with dimensions 6-ft L × 4-in W × 10-in H, is made from a wooden frame with clear acrylic sheets as sides. First, 1/8-in thick grooves are cut into the bottom and ends of the wooden frame. (*Continues.*)

in-class simulations. Once a wave is generated, it moves quickly across the tank so we suggest building a tank as long as possible—six feet worked very well for us and we suspect the phenomena we are trying to observe would be difficult to see in a tank shorter than four feet. With a little practice, one can generate nice impulsive disturbances and clearly see the amplitude, wavelength and speed of the wave change as it propagates. The complete model is shown in Figures 7.4a, b, and c.

7.1.5.2 Simulations

We devoted one 50-minute class to pilot our activity. We spent half of the class establishing that tsunamis are *SWW*s and the remainder of the class running simulations while eliciting student responses and answering questions. We did not ask students to complete any problems or homework assignments. Our class discussion simply focused on energy considerations. The speed of a *SWW* is proportional to $\sqrt{g \cdot d}$, where d is the water depth, while the transported energy depends on both the wave's speed and amplitude (this result is proved in the final section of this chapter). As a tsunami enters the shallow water near a coast and d gets smaller, its velocity slows and wavelength decreases. Thus, in order to conserve energy, the amplitude of a tsunami dramatically grows! A tsunami that was only a few feet in amplitude in the deep ocean and essentially imperceptible, can grow in amplitude to devastating heights as it moves to shallow water. This increase in amplitude with decreasing water depth is the dramatic effect we see on

Figure 7.4: (*Continued.*) (b) Construction of the *"tsunami tank."* Next, 1/8-in thick clear acrylic sheets are inserted into the grooves and cemented into place using construction adhesive.

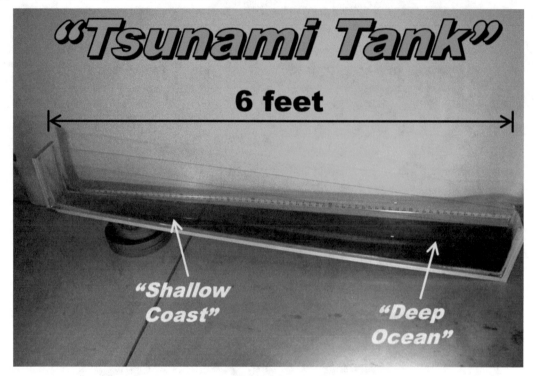

Figure 7.4: (*Continued.*) (c) The completed *"tsunami tank."* To simulate an ocean with varying water-depths, the tank is tipped and then filled with water so that the water-depth linearly changes from a *"deep ocean"* to a *"shallow coast."* An impulsive disturbance in the deep portion of the tank simulates an earthquake that in turn generates a tsunami-like wave.

newscasts. Figures 7.5a–j depict typical results from our simulation and clearly illustrate these phenomena. During the in-class simulations, the most striking effects occur as the wave climbs the shore and encounters a rapidly decreasing water depth. Although reductions in velocity and wavelength are easy to observe, students react most strongly to the very noticeable increase in amplitude.

7.1.5.3   Conclusions

We set out to create a simulation of tsunamis for introductory physics classes. Although highly sophisticated models and animations already exist, constructing these models involves advanced concepts from physics while the animations require highly specialized algorithms and powerful computers. By creating a *"tsunami tank,"* teachers of introductory physics can produce actual in-class simulations that capture the essential physics of tsunamis and make this timely topic appropriate for their classrooms.

Figure 7.5: Images (a)–(e) show the simulated tsunami traveling in the deep portion of the tank. Notice that the bottom of the tank is not visible in these images. Images (f)–(j) show the *same wave* as it travels in the shallow portion of the tank, where the bottom can be seen sloping upward in the bottom-left corner of each image. As the wave moves into the shallow water-depth, comparing images (c) and (h) best demonstrates the dramatic increase in amplitude (notice the height and shape of the wave in image (h) vs. image (c)) while comparing images (d) and (i) best demonstrates the decrease in both wavelength and velocity.

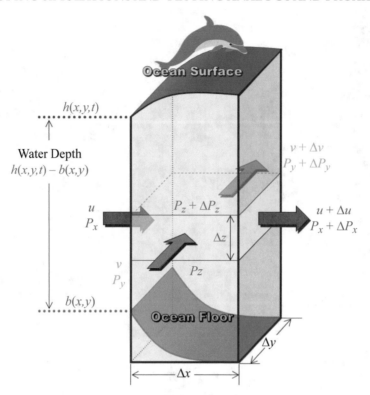

Figure 7.6: Sample Volume of Water in Hydrostatic Equilibrium. The ocean is in a state of *"hydrostatic equilibrium."* Physically, the weight of any infinitesimally small volume of water, $\Delta V = \Delta x \cdot \Delta y \cdot \Delta z$, is in balance with the pressure-gradient force across the top and bottom faces. This leads to a pressure-gradient given by the expression: $-\Delta P_z = \rho \cdot g \cdot \Delta z$ or $\frac{\partial P}{\partial z} = -\rho \cdot g$. Clipart of dolphin, courtesy of Wikimedia.

7.1.6   SIMULATION 2 – COMPUTER-BASED MODEL

We now turn our attention to the question: *"Can the SWW equations be derived in an introductory physics class?"* We present the brief (somewhat) derivation of the *SWW* equations that we used in class and direct the interested reader to our website for a more detailed derivation. However, we stress that our primary goal was to expose students to more advanced topics rather than to engage them in a detailed description of fluid dynamics and water wave models. Figure 7.6 depicts the ocean floor defined by a specified time-independent profile $b(x, y)$. We seek to compute and animate a tsunami's time-dependent surface profile $h(x, y, t)$. In order to accomplish this, we derived the *SWW* equations using four steps.

7.1.6.1   Step 1: The Shallow Water Wave Assumption

We start the derivation by following an infinitesimally small *"fluid particle"* of mass Δm, confined in the volume of dimensions $\Delta V = \Delta x \cdot \Delta y \cdot \Delta z$, as shown in Figure 7.6. We apply Newton's Second Law to a *"fluid particle"* located at the point $(x(t), y(t), z(t))$ parameterized by time t, since the particle's location changes in time. We define the particle to be an infinitesimally small mass of fluid Δm, confined in a volume of dimensions $\Delta x \cdot \Delta y \cdot \Delta z$.

Newton's Second Law relates the sum of the forces applied to this volume at an instant in time t to its mass and velocity: $\mathbf{F}(t) = \Delta m \cdot \frac{d}{dt} \mathbf{V}(t)$. Based on our initial assumptions, the only forces acting on the mass are the pressure-gradient forces, ΔP_x, ΔP_y, and ΔP_z, across the six sides of the volume and the force of gravity along the vertical axis:

$$\Delta m \frac{d\mathbf{V}}{dt} = - \left(\underbrace{\Delta P_x \cdot \Delta y \cdot \Delta z}_{F_x}, \; \underbrace{\Delta P_y \cdot \Delta x \cdot \Delta z}_{F_y}, \; \underbrace{\Delta P_z \cdot \Delta x \cdot \Delta y \; + \; \rho \cdot g \cdot \Delta x \cdot \Delta y \cdot \Delta z}_{F_z} \right).$$

The negative sign indicates that the direction of the pressure-gradient force is opposite to the direction of the pressure change. Since the pressure P depends on both position and time, we can write $P = P(x, y, z, t)$ and introduce the concept of a partial derivative to the class as simply the rate of change, (i.e., the slope) of P in a direction parallel to one of the coordinate axes, allowing us to make the approximations $\Delta P_x = \left(\frac{\partial P}{\partial x} \right) \Delta x$, $\Delta P_y = \left(\frac{\partial P}{\partial y} \right) \Delta y$, and $\Delta P_z = \left(\frac{\partial P}{\partial z} \right) \Delta z$. These approximations give:

$$\Delta m \frac{d\mathbf{V}}{dt} = - \left(\frac{\partial P}{\partial x}, \; \frac{\partial P}{\partial y}, \; \frac{\partial P}{\partial z} \; + \; \rho \cdot g \right) \Delta V,$$

where $\Delta V = \Delta x \cdot \Delta y \cdot \Delta z$ is the infinitesimal volume element.

7.1.6.2   Step 2: The Resulting Pressure Gradient

We now make use of the second of our three equivalent *SWW* assumptions, (i.e., recall our demonstration with the cookie tray), namely, that the water is in hydrostatic equilibrium. This assumption states that no net force is exerted on our particle along the vertical axis; therefore, the z-component of our force equation is zero, requiring $\frac{\partial P}{\partial z} = -\rho \cdot g$, and leaving us:

$$\frac{d\mathbf{V}}{dt} = -\frac{1}{\rho} \left(\frac{\partial P}{\partial x}, \; \frac{\partial P}{\partial y}, \; 0 \right), \tag{7.1}$$

where we have used $\rho = \frac{\Delta m}{\Delta V}$.

Using $\frac{\partial P}{\partial z} = -\rho \cdot g$, we derive an expression for the pressure at the location of our particle $(x(t), y(t), z(t))$. The pressure at this point is equal to the weight of the water above it:

$$P(x,y,z,t)|_{z=h(x,y,t)} - P(x,y,z,t) = \int_z^{h(x,y,t)} \frac{\partial P}{\partial z} dz'$$

$$= -\int_z^{h(x,y,t)} \rho \cdot g \cdot dz' = -\rho \cdot g \cdot [h(x,y,t) - z],$$

or

$$P(x,y,z,t) = \rho \cdot g \cdot [h(x,y,t) - z], \tag{7.2}$$

where we have assumed $P(x,y,z,t)|_{z=h(x,y,t)} = 0$. The result is the standard linear pressure expression derived in most introductory physics textbooks.

7.1.6.3   Step 3: The Convective Derivative ("The Momentum Equations")

The final step in the application of Newton's Second Law is to transform the time derivative on the left-hand side of Eq. (7.1) into a sum of partial derivatives [9]. We again use the notion of a partial derivative as a slope to argue that a small change in time Δt and small changes in the Cartesian position coordinates Δx and Δy produce a small change in velocity $\Delta \mathbf{V}$ given by:

$$\Delta \mathbf{V} = \left(\frac{\partial \mathbf{V}}{\partial t} \cdot \Delta t \right) + \left(\frac{\partial \mathbf{V}}{\partial x} \cdot \Delta x \right) + \left(\frac{\partial \mathbf{V}}{\partial y} \cdot \Delta y \right).$$

Dividing by Δt and passing to the limit, we obtain the rate of change,

$$\frac{d\mathbf{V}}{dt} = \frac{\partial \mathbf{V}}{\partial t} + \left(u \cdot \frac{\partial \mathbf{V}}{\partial x} \right) + \left(v \cdot \frac{\partial \mathbf{V}}{\partial y} \right), \tag{7.3}$$

where $u = \frac{dx}{dt}$ and $v = \frac{dy}{dt}$ are the depth-averaged horizontal velocity components discussed earlier.

Combining Eqs. (7.1), (7.2), and (7.3), and separating the x- and y-components, we obtain two of the three *SWW* equations:

x-Momentum Equation

$$\frac{\partial u}{\partial t} + \left(u \cdot \frac{\partial u}{\partial x} \right) + \left(v \cdot \frac{\partial u}{\partial y} \right) + \left(g \cdot \frac{\partial h}{\partial x} \right) = 0, \tag{7.4}$$

y-Momentum Equation

$$\underbrace{\frac{\partial v}{\partial t}}_{(1)} + \underbrace{\left(u \cdot \frac{\partial v}{\partial x} \right) + \left(v \cdot \frac{\partial v}{\partial y} \right)}_{(2)} + \underbrace{\left(g \cdot \frac{\partial h}{\partial y} \right)}_{(3)} = 0. \tag{7.5}$$

(1) *Convective Derivative*: these terms show the two ways in which the velocity of a "fluid particle" can change—it can change because the whole velocity field is changing (a process present even if the particle itself is at rest) or …

(2) …it can change by moving to a position where the velocity is different (a process present even if the velocity field as a whole is steady) [3].

(3) *External Forces*: this term represents the horizontal pressure gradient determined by the weight of the water above the "fluid particle."

Equations (7.4) and (7.5) are called the SWW "Momentum Equations."

7.1.6.4 Step 4: "The Mass Continuity Equation"

With three unknown functions $u(x, y, t)$, $v(x, y, t)$, and $h(x, y, t)$, and only two equations of constraint, we need one additional constraint to complete our derivation of the *SWW* equations. This constraint is derived from the principle of Conservation of Mass applied to the rectangular column of water extending from the ocean surface to the ocean floor as depicted in Figure 7.6. The principle states that the rate at which mass enters the column (in $\frac{\text{kg}}{\text{s}}$) minus the rate at which mass leaves the column must equal the rate of change of mass within the column. For convenience, let $H = h - b$. Conservation of Mass gives us:

$$\text{Mass}_{\text{rate of change}} = (\text{Mass-flow}_{\text{out}} - \text{Mass-flow}_{\text{in}}),$$

where

$$\text{Mass-flow}_{\text{in}} = (\text{Mass-flow}_{\text{across left face}} + \text{Mass-flow}_{\text{across front face}})$$
$$= \rho(u \cdot H)\Delta y + \rho(v \cdot H)\Delta x,$$

$$\text{Mass-flow}_{\text{out}} = (\text{Mass-flow}_{\text{across right face}} + \text{Mass-flow}_{\text{across back face}})$$
$$= \rho[(u + \Delta u) \cdot (H + \Delta H)]\Delta y + \rho[(v + \Delta v) \cdot (H + \Delta H)]\Delta x, \text{ and}$$
$$\text{Mass}_{\text{rate of change}} = -\rho \left(\frac{\partial h}{\partial t} \right) \Delta x \cdot \Delta y.$$

In the first two equations, we have made use of the assumption that u and v are depth-averaged velocity fields that do not vary along the vertical axis. In the third equation, we have used the ocean floor's topology time-independence. Ignoring terms of the order $\Delta H \cdot \Delta x$ and $\Delta H \cdot \Delta y$, and passing to the limit, we get the last of the *SWW* equations:

Mass Continuity Equation:

$$\underbrace{\frac{\partial h}{\partial t}}_{(1)} + \underbrace{\frac{\partial}{\partial x}[u \cdot (h - b)] + \frac{\partial}{\partial y}[v \cdot (h - b)]}_{(2)} = 0. \tag{7.6}$$

(1) *Mass Rate of Change*: This term represents the mass change in the water column.

(2): *Change of Mass*: These terms represent the mass entering and exiting across the four sides of the water column.

Equation (7.6) is known as the *SWW "Mass Continuity Equation."* Collectively, Eqs. (7.4), (7.5), and (7.6) are known as the *"Shallow Water Wave Equations."*

7.1.7   COMPUTER MODELING

We first used a standard finite-difference technique to solve the *SWW* equations [10]. In this technique the continuous domain functions u, v, and h are transformed into discrete domain functions: for example, $u(x, y, t) = u_{i,j}^n$, where i and j are the space indices and n is the time index. This formulation allows partial derivatives to be expressed in finite difference form: for example, $\frac{\partial u}{\partial x} \approx \frac{\Delta u}{\Delta x} = \frac{u_{i+1,j}^n - u_{i,j}^n}{\Delta x}$. In order to animate a tsunami's motion, at each time-step $n + 1$ we displayed its surface profile $h_{i,j}^{n+1}$ for all i and j. As a second method, we employed *Mathematica*, a commonly used symbolic mathematics package. *Mathematica* has built-in functions for numerically solving systems of partial differential equations, given the appropriate initial- and boundary-conditions, as well as built-in functions for displaying 3-dimensional graphs of the results.

Specifically, we modeled the Indian Ocean between Africa and Indonesia as an infinitely long trough with a width of 7,000 km and a depth of 7,000 m. We created a piecewise continuous function to form the ocean floor: extending ~1,500 km from each coast was a shelf that linearly lowered the ocean floor from sea-level to a depth of 1,000 m. This shelf was included to more dramatically demonstrate the variation in a tsunami's amplitude as it moves closer to shore. From the edge of each of these shelves, a cosine function then smoothly lowered the ocean floor to a flat bottom at a depth of 7,000 m. Finally, we placed a 3,600-m high Gaussian-ridge along the center-line of our trough to approximate the Mid-Indian Ocean ridge.

To simulate the initial volume of displaced water, we set the initial condition of the ocean surface to be a "Gaussian pluck," ~1,200 km wide, centered to the right of our Mid-Indian Ocean ridge to roughly coincide with the epicenter of the 2004 earthquake. Since we would not resolve a 15 m displacement on a vertical scale of 7,000 m, we set the initial Gaussian-pluck to be 750-m high and argued that similar changes in amplitude would occur for the 15-m high displacement. The complete model is depicted in Figure 7.7.

7.1.8   RESULTS

Figure 7.8 and Table 7.1 depict typical results from our finite-difference simulations and clearly illustrates these phenomena. As our simulated tsunami propagates across the ocean, any reduction in water depth $(h - b)$ produces a corresponding increase in the tsunami's amplitude and decrease in its velocity. The most striking effects occur when the tsunami passes over the Mid-Indian Ocean ridge or as it enters the coastal shelves. The wave that moves toward Africa almost

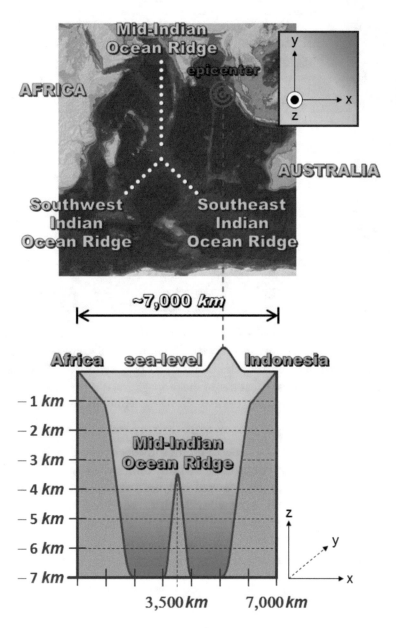

Figure 7.7: Bathymetric plot of the Indian Ocean and corresponding simulated profile of the ocean floor [5].

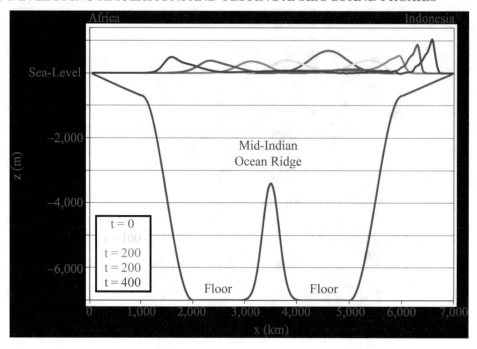

Figure 7.8: Simulated propagation of a Tsunami in the Indian Ocean. Time, t is a normalized time and A_o is the tsunami's amplitude at $t = 0$. The initial displacement, we imagine to have been created by an underwater earthquake, is modeled by a symmetric Gaussian pluck. The simulation displays the characteristic features of the propagation of a shallow water wave, namely, as the water depth decreases, the wave's velocity decreases while its amplitude increases.

immediately encounters this ridge so its speed significantly decreases while its amplitude slightly grows. Once past the ridge, the wave encounters a fairly uniform water depth until the very end of the simulation. The wave that moves toward Indonesia shows even more dramatic effects. This wave encounters a rapidly decreasing water depth so reductions in velocity and growths in amplitude are easy to observe. We stop the simulation before the tsunami reaches shore because our model breaks down when frictional forces, higher amplitudes, and a decreasing wavelength-to-depth ratio compromise the validity of our assumptions.

7.2   CONCLUSIONS

We set out to create a mathematical model and animation of a tsunami for introductory physics classes. Although highly sophisticated models and animations already exist, constructing these models involves advanced concepts from physics while the animations require highly specialized algorithms and powerful computers. By limiting our model to a tsunami in the deep ocean,

Table 7.1: Results from our finite-difference simulations

	Wave Moving Left (toward Africa)				Wave Moving Right (toward Indonesia)			
	Depth (m)	x (km)	v (arbitrary)	A/A$_o$	Depth (m)	x (km)	v (arbitrary)	A/A$_o$
t = 0	7,000	4,550	0	1	7,000	4,550	0	1
t = 50	7,000	4,340	80	0.958	7,000	4,900	73	1.056
t = 100	6,274	3,780	40	0.965	4,697	5,411	50	1.058
t = 150	3,402	3,500	40	0.965	2,052	5,691	40	1.068
t = 200	7,000	3,129	54	0.958	783	5,929	34	1.077
t = 250	7,000	2,709	54	0.958	621	6,111	27	1.090
t = 300	7,000	2,331	54	0.958	490	6,300	26	1.125
t = 350	6,975	1,960	53	0.965	392	6,440	20	1.140
t = 400	4,696	1,589	50	0.975	294	6,580	20	1.150

we can successfully approximate the problem as a *"Shallow Water Wave."* Therefore, teachers of introductory physics can construct a model that captures the essential physics of a tsunami by making certain assumptions that render the problem appropriate to their students.

7.2.1 PROOF: THE VELOCITY OF A SHALLOW WATER WAVE

To close this chapter, we tie-up an annoying loose ends by proving mathematically that the velocity of a *"Shallow Water Wave"* is proportional to the square-root of the water-depth. The *"Shallow Water Wave"* (SWW) equations include the two *"Momentum"* equations and the *"Mass Continuity"* equation:

$$\textbf{x-Momentum:} \quad \frac{\partial u}{\partial t} + \left(u \cdot \frac{\partial u}{\partial x} \right) + \left(v \cdot \frac{\partial u}{\partial y} \right) + \left(g \cdot \frac{\partial h}{\partial x} \right) = 0, \qquad (7.4)$$

$$\textbf{y-Momentum:} \quad \frac{\partial v}{\partial t} + \left(u \cdot \frac{\partial v}{\partial x} \right) + \left(v \cdot \frac{\partial v}{\partial y} \right) + \left(g \cdot \frac{\partial h}{\partial y} \right) = 0, \qquad (7.5)$$

and

$$\textbf{Mass Continuity Equation:} \quad \frac{\partial h}{\partial t} + \frac{\partial}{\partial x} \left[u \cdot (h - b) \right] + \frac{\partial}{\partial y} \left[v \cdot (h - b) \right] = 0. \qquad (7.6)$$

Let us assume that the ocean is in a basic state of equilibrium in which $\vec{v} = \vec{v}_o = 0$ and $h = h_o = $ constant. We now assume that the state of the ocean can be represented by small perturbations about this equilibrium state. We therefore assume that $h = h_o (1 + \eta)$, where $|\eta| \ll 1$, and quadratic terms in $\vec{v}$ and η are so small that they can be ignored. The *SWW* equations reduce to:

$$\text{x-Momentum:}\quad \frac{\partial u}{\partial t} + \left(g \cdot h_o \cdot \frac{\partial \eta}{\partial x} \right) = 0, \tag{7.4}$$

$$\text{y-Momentum:}\quad \frac{\partial v}{\partial t} + \left(g \cdot h_o \cdot \frac{\partial \eta}{\partial y} \right) = 0, \tag{7.5}$$

and

$$\text{Mass Continuity Equation:}\quad \frac{\partial \eta}{\partial t} + \frac{\partial u}{\partial x} + \frac{\partial v}{\partial y} = 0. \tag{7.6}$$

We now assume that $\vec{v}$ and η are proportional to $\exp\left[i\left(k_x \cdot x + k_y \cdot y - \omega \cdot t\right)\right]$, where k is the wavenumber, ω is the frequency, and $c = \frac{\omega}{k}$ is the speed with which the waveform, not the fluid, propagates. The *SWW* equations now reduce to:

$$\text{x-Momentum:} - \omega u + (g \cdot h_o \cdot k_x \cdot \eta) = 0, \tag{7.7}$$
$$\text{y-Momentum:} - \omega v + \left(g \cdot h_o \cdot k_y \cdot \eta\right) = 0, \tag{7.8}$$

and

$$\text{Mass Continuity Equation:} - \omega \eta + (k_x \cdot u) + \left(k_y \cdot v\right) = 0. \tag{7.9}$$

For convenience, we will write this system of three linear equations in matrix form:

$$\begin{pmatrix} g \cdot h_o \cdot k_x & -\omega & 0 \\ g \cdot h_o \cdot k_y & 0 & -\omega \\ -\omega & k_x & k_y \end{pmatrix} \begin{pmatrix} \eta \\ u \\ v \end{pmatrix} = 0. \tag{7.10}$$

Equation (7.10) is most easily solved by setting the determinant equal to zero. The resulting expression for ω is:

$$\omega^2 = g \cdot h_o \cdot \left(k_x^2 + k_y^2\right) = g \cdot h_o \cdot k^2, \tag{7.11}$$

where $k^2 = k_x^2 + k_y^2$. Finally, Eq. (7.11) gives us the velocity of the waveform and demonstrates the relation between the velocity of a *SWW* and the water depth:

$$c = \frac{\omega}{k} = \sqrt{g \cdot h_o}. \tag{7.12}$$

7.3   ACKNOWLEDGMENTS

The authors wish to acknowledge Jay Tarby, Tom Philips, and Dick Bradley of the *John Carroll University Faculty Technology Innovation Center* for the use of video equipment and for digitizing our images. We also acknowledge *DJ Glass and Mirror* for their many helpful suggestions in constructing the *"tsunami tank."*

7.4   REFERENCES

[1] Photograph courtesy of Wikimedia and David Redevik (RELEASED). This photograph may be used by anyone for any purpose. 98

[2] G. A. DiLisi and R. A. Rarick, Modeling the 2004 Indian Ocean tsunami for introductory physics students, *Phys. Teach.*, 44:585–588, December 2006. DOI: 10.1119/1.2396776. 98

[3] D. H. Peregrine, Equations for water waves and the approximations behind them, *Waves on Beaches and Resulting Sediment Transport*, R. Meyer, Ed., Academic Press, New York, 1972. DOI: 10.1016/b978-0-12-493250-0.50007-2. 100, 111

[4] D. T. Sandwell, et al., *Bathymetry from Space: White Paper in Support of a High Resolution, Ocean Altimeter Mission*, Scripps Institute of Oceanography, La Jolla, CA. 101

[5] Bathymetric plot courtesy of Wikimedia. Permission granted under terms of the GNU Free Documentation License. 102, 113

[6] The United States Geological Survey, USGS Home Page: Historic Worldwide Earthquakes. http://earthquake.usgs.gov/regional/world/historical.php 101

[7] Photograph courtesy of Wikimedia. Public Domain. 103

[8] In the 1830s, the Scottish naval engineer John Scott Russell built water troughs, 30 feet in length, in his garden and studied the large waves created by either dropping masses into, or removing masses from, one end of the trough. In 1834, Russell was the first to observe and describe the *Wave of Translation*, which today is known as a *Soliton*, or *Russell Solitary Wave*. See P. G. Drazin and R. S. Johnson, *Solitons: An Introduction*, Cambridge Texts in Applied Mathematics, Cambridge University Press, 1989. 103

[9] D. J. Tritton, *Physical Fluid Dynamics*, Van Nostrand Reinhold Co. Ltd., UK, 1977. DOI: 10.1007/978-94-009-9992-3. 110

[10] N. J. Giordano, *Computational Physics*, Prentice Hall, Upper Saddle River, NJ, 1997. 112

CHAPTER 8

Incorporating Active Areas of Research and Asking Complex Questions

Case studies allow teachers to discuss active areas of research and ask complex questions while still covering the appropriate content and maintaining the appropriate level

Teachers are often handcuffed to teach predetermined, district-, or department-prescribed curricula. Case studies allow front-line research to be brought into the classroom. Teachers can emphasize to students the importance of staying current with recent developments in scientific research and demonstrate how this informs teaching. The associated analyses can potentially showcase real-world applications of the material that is being taught and show students that the answers to complex questions are not always known and sometimes remain a mystery. The sinking of the "unsinkable" R.M.S. Titanic on its maiden trans-Atlantic voyage is the hallmark example of such an opportunity. On April 14, 1912, the massive British passenger liner struck an iceberg and sank in under three hours. Over 1,500 passengers and crew perished. We present the historical debate as: *"Why did Titanic hit the iceberg in the first place?"* [1].

8.1 THE SINKING OF THE R.M.S. TITANIC: A CASE STUDY IN THERMAL INVERSION AND ATMOSPHERIC REFRACTION PHENOMENA

Gregory A. DiLisi, *John Carroll University, University Heights, Ohio*

8.1.1 INTRODUCTION

On April 14, 1912 the British passenger liner R.M.S. Titanic struck an iceberg [1]. The ship sank in a fraction of the time designers had estimated following a worst case scenario. The purpose of this article is to examine the atmospheric refractive phenomena that might have played a significant role in obscuring the iceberg from Titanic's two lookouts. We describe a way in which these phenomena can be easily and inexpensively brought to students in our introductory physics classrooms.

Figure 8.1: The iceberg thought to have collided with the Titanic. This photograph was taken on the morning of April 15, 1912 by the chief steward aboard the German ocean liner, S. S. Prinz Adalbert [2]. The steward noted that a streak of red paint, allegedly from Titanic's hull, was "plainly visible" along the waterline of the iceberg [3].

8.1.2 THE R.M.S. TITANIC

On April 10, 1912, the R.M.S. Titanic left Southampton, England carrying 920 passengers. Under the command of Captain Edward Smith, the ship stopped in Cherbourg, France where an additional 274 passengers boarded and 24 passengers disembarked. After another brief stop in Queenstown, Ireland, where another 123 passengers boarded, the ship started its maiden trans-Atlantic voyage to New York Harbor. As Titanic approached Newfoundland, several ships issued warnings of icebergs and drifting ice in the area. By April 14, the ship had received at least six such warnings. As was standard maritime practice at the time, Captain Smith regarded these ice warnings as cautionary advisories and continued to sail at 22 knots. Titanic's top speed was 23 knots (25 mph). At 11:40 pm (ship's time), lookout Frederick Fleet, stationed on the port-side of Titanic's crow's nest, spotted an iceberg 1/4-mile dead-ahead of the ship (see Figure 8.1). Fleet phoned the bridge and issued his famous warning: "Iceberg, right ahead!" [4]. According to the second lookout, Reginald Lee, stationed on the starboard-side of the crow's nest: "*It was a dark mass that came through that haze and there was no white appearing until it was just close alongside the ship, and that was just a fringe at the top*" [5].

Officers responded quickly. They slowed the massive liner and tried to steer it around the iceberg. First Officer William Murdoch turned Titanic to port to clear the bow from collision then immediately turned to starboard to clear the stern. Murdoch's efforts failed. Titanic's starboard-bow hit the iceberg below the waterline with a glancing blow. The iceberg did not tear into Titanic's hull, as is sometimes portrayed; instead, it merely dented the ship. Seams in the hull buckled, separated, and gaped. Water poured in. Titanic's designers had built the ship for a worst case scenario of a head-on collision with another ship, resulting in the flooding of four of its watertight compartments. With four compartments flooded, Titanic could stay afloat for 2–3 days, giving rescue ships plenty of time to arrive [6]. However, the iceberg had damaged 300 feet of the hull, breaching six of the watertight compartments. Titanic was doomed! Only 2 hours and 40 minutes later, around 2:20 am on the morning of April 15, the largest ocean liner in service at the time disappeared beneath the surface of the Atlantic Ocean. Titanic now lies broken in two, 2.4 miles underwater, 370 miles southeast of the tip of Newfoundland and over 1,000 miles from its final destination. Of the estimated 2,227 passengers and crew, 710 (32%) survived while 1,517 (832 passengers and 685 crew) perished [7]. Given the number of lifeboats available on the ship, only 53% of personnel could have survived.

8.1.3   STATEMENT OF THE PROBLEM

The sinking of the Titanic is the subject of many documentaries, books, movies, and conspiracy theories. Probably no tragedy has been the subject of more forensic re-examination than Titanic. Analyses generally fall into two camps, the first of which attempts to answer the question: "Why did Titanic sink after only a minor collision with an iceberg?" Theories typically focus on design flaws and material failures: the rudder was too small to turn the ship properly ... the watertight compartments were not tall enough ... the low-grade steel of Titanic's hull cracked, rather than deformed, under the pressures and temperatures encountered during the disaster (a phenomenon known as "brittle fracturing") ... the heads on some of the wrought iron rivets that held Titanic's hull together broke under the force of the collision, resulting in whole sections of the ship coming apart. A recent theory even examined the idea that a smoldering fire burned in one of Titanic's boiler rooms as a result of spontaneous combustion underneath a huge pile of coal. Supposedly, this fire weakened Titanic's hull, making the ship susceptible to the impact with the iceberg. The second camp attempts to answer the question: "Why did so many passengers perish after the collision?" What if Titanic had carried its full complement of lifeboats, instead of the mere 20 that it carried? Why were so many lifeboats launched only partially loaded? (Fully loaded, the lifeboats could have saved as many as 1,178 passengers and crew.) Why were passengers so reluctant to leave the ship in the early stages of the evacuation? Even after observing Titanic's distress rockets, why did the crew of the S.S. Californian, the closest ship to Titanic, not render aid?

Fascinating as these discussions are, a more fundamental question remains seldom explored: "Why did Titanic hit the iceberg in the first place?" Certainly, several factors contributed to the inability of the two lookouts to see the on-coming iceberg: neither lookout had binoculars; the night was moonless; and the sea was calm so waves were not lapping against the iceberg making it easy to spot [8]. However, records indicate that 1912 was an especially dangerous year to sail and that the lookouts should have been on higher-than-normal alert for drifting ice. In that year, a total of 1,038 icebergs entered the North Atlantic shipping lanes before melting away, with nearly 400 crossing into the lanes in April alone [9]. During an average year, only a few hundred icebergs typically enter the lanes. Given this data and having received at least six warnings of drifting ice in the area, one might think that two highly trained lookouts would have more-readily spotted the iceberg in time to avert the disaster. What could have obscured the iceberg from their view?

8.1.4   THERMAL INVERSION

The key to understanding the refractive phenomena which may have hidden the iceberg from Titanic's lookouts is to examine the typical atmospheric conditions near the location where the massive passenger liner sank. Titanic's wreckage lies at 41°43′37.0″N, 49°56′53.7W, right along the edge of the Grand Banks of Newfoundland, a series of shallow, underwater plateaus located at the south-east tip of Newfoundland. Here, a major North Atlantic Ocean surface current, called the Labrador Current, brings cold water from the Arctic Ocean, along the western edge of the Labrador Sea, into the Grand Banks area to mix with the warm waters of the Gulf Stream. The mixing of these waters combines with the underwater plateaus to produce some of the most nutrient-rich fishing grounds in the world. However, these conditions also create a host of fascinating atmospheric phenomena that wreak havoc on maritime navigation. In 2012, National Geographic produced a documentary titled: "Titanic: Case Closed," in which historian Tim Maltin, with help from astronomer Andrew Young, advanced the theory that atmospheric conditions in the Grand Banks region all but cloaked the iceberg from Titanic's lookouts. Maltin published his findings in a book, appropriately titled: *Titanic: A Very Deceiving Night* [5]. In a nutshell, Maltin's theory is that a flip in the thermal gradient of earth's lower atmosphere caused refractive phenomena that concealed the iceberg from the eyes of Titanic's lookouts.

Normally, air temperature in the lower atmosphere steadily decreases with elevation. A good rule of thumb for the tropospheric temperature gradient is −6.5 K/km (meteorologists use the terminology: "Lapse rate of +6.5 K/km"). Humans are accustomed to the bending of light occurring in this atmospheric profile. In fact, as teachers of introductory physics, we often discuss the refractive phenomena associated with such a profile. For example, on sunny days, when the air near the surface of a road becomes even more heated and less optically dense than usual, light from distant objects drastically curves *upward* toward the cooler air above, *opposite of the curvature of the earth*, resulting in the appearance of inverted images that appear *below* the actual objects. We describe these images as "inferior mirages," using the word "inferior" to indicate that an observer sees the inverted image *below* the actual object. Another example of this phenomenon is the classic desert mirage. Mirages are not optical illusions—they are actual images caused by rays of light bending and entering the eye of an observer. Optical illusions, on the other hand, are misinterpretations of visual information by the human brain. Of course, mirages can trigger optical illusions and trick observers into seeing something that does not

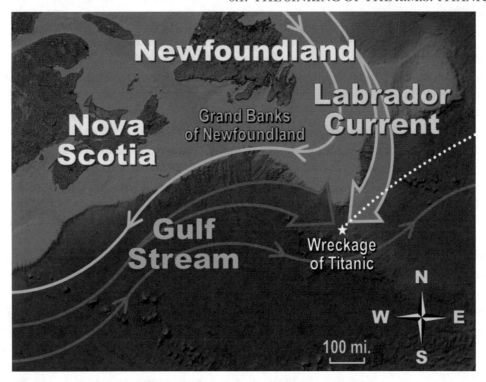

Figure 8.2: The area of the Grand Banks of Newfoundland. The Grand Banks of Newfoundland lie south-east of the tip of Newfoundland. The wreckage of the R.M.S. Titanic rests where the cold waters of Labrador Current mix with the warm waters of the Gulf Stream. Map courtesy of Google Maps.

exist. A simple way to distinguish between mirages and optical illusions is that mirages can be photographed, optical illusions cannot.

As Titanic approached Newfoundland, it entered a region with a much different atmospheric profile—a region where humans are not accustomed to the accompanying refractive phenomena. Specifically, as Titanic sailed from the warm waters of the Gulf Stream into the cold waters of the Labrador Current, a cool, optically dense layer of air engulfed the ship while warmer air hovered above. This flipped thermal profile is known as a "thermal inversion" or "inversion layer." Thermal inversions often occur at night and over large bodies of cold water, especially near the boundaries of cold and warm waters or near coastlines. The thermal inversion encountered by Titanic was acknowledged by Second Officer, Charles Lightoller, who testified that on April 14, the air temperature dropped 10°C, to almost freezing, between 7 pm and 9 pm [5].

8.1.5   REFRACTIVE PHENOMENA

In a thermally inverted atmosphere, some very interesting and visually stunning refractive phenomena occur. For example, light from distant objects can drastically curve *downward, in the same direction as the curvature of the earth*, resulting in the appearance of inverted images that appear *above* the actual objects. We describe these images as "superior mirages," using the word "superior" to indicate that an observer sees the inverted image *above* the actual object. Again, a mirage is an actual, inverted image, not an optical illusion. Thus, an object like an on-coming boat or iceberg may appear higher in the sky than the object actually is. *Could a superior mirage have played tricks on the eyes of Titanic's lookouts before the ship struck the iceberg?* As strange and disorienting as superior mirages may be, another visually-perplexing phenomenon may have been even more responsible for obscuring the iceberg from Titanic's lookouts—the phenomenon of "looming." Looming occurs when the bending of light from distant objects becomes comparable to the curvature of the earth and objects *beyond the horizon* actually become visible to an observer. Objects like the ocean's surface or a distant coast do not become distorted, they simply become vertically displaced. On the night of Titanic's fatal encounter, the thermal inversion caused by the mixing of the Labrador Current and Gulf Stream created the perfect conditions for the looming of the distant sea. With the iceberg silhouetted against a higher "false horizon," it was essentially cloaked from the eyes of Titanic's lookouts [10]. In a nutshell, the dimly lit iceberg was concealed against the backdrop of a dark, false horizon that was vertically displaced by looming in a thermally-inverted atmosphere (see Figure 8.3).

   Other effects like "ducting" (the trapping of light rays in certain layers of the atmosphere), "sinking" (the disappearance of objects below the horizon which are usually visible), "towering" (the apparent vertical stretching of an object), and "stooping" (the apparent vertical compression of an object), as well as combinations of these effects, are possible given the appropriate atmospheric profile. Of course, these refractive phenomena are highly sensitive to the thickness, uniformity, and smoothness of transition of the temperature gradient within the atmosphere. Also, these effects depend strongly on the height of the observer and the distance between the observer and the object.

8.1.6   CLASSROOM DEMONSTRATION

A discussion of thermal inversions and their accompanying refractive phenomena fit perfectly into an introductory physics sequence. These topics have a home in any course that addresses optics, refraction, or mirages. Although we would like to create an in-class demonstration of *looming*, this phenomenon technically occurs as a result of the bending of light rays from beyond the horizon. A simulation of looming would necessitate an awfully large classroom and apparatus! Instead, we describe a simple demonstration of *superior mirages*, noting that the bending of light is similar to that exhibited during looming. Our demonstration is inexpensive, easy to set-up, and results in visually-powerful effects. The images presented in this article are indeed representative of the effects seen in class but do not do justice to seeing them in person.

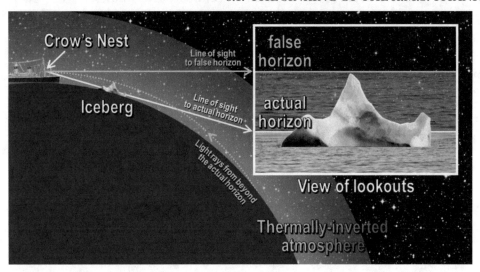

Figure 8.3: Looming. Light rays (green-dotted line) from the sea beyond the actual horizon (yellow line) bend and enter the eyes of the lookouts stationed in Titanic's crow's nest. The distant sea appears higher in the sky than it really is. The lookouts see the distant sea as a "false horizon" (green solid line), against which the iceberg is concealed. Thermal inversions can have deadly consequences for sea travelers as humans are unaccustomed to the accompanying refractive phenomena.

A ten-gallon fish tank, measuring $10''L \times 20''W \times 12''H$, was purchased at a local pet shop (our entire $30-investment in this demonstration was on the tank). At one end of the tank, we placed a bright yellow trifold board. A sheet of black construction paper was glued to the trifold to simulate a horizon. At the other end of the tank, we mounted a camera to a vertical translation stage. The translation stage allowed us to slightly raise and lower the camera while eliminating any effects of parallax due to tilting of the camera. The camera was connected to the classroom projection system so that students could get an up-close view of the phenomenon occurring across the tank. Between the camera and tank, we placed a simulated iceberg that we shaped using modeling clay. The entire set-up is shown in Figure 8.4.

Next, we filled the tank with solutions to simulate a thermally-inverted atmosphere. We strongly advise filling the tank in the classroom where the demonstration is to take place—the tank cannot be filled elsewhere and transported to class without disrupting the delicate layering of solutions needed for the demonstration. Our demonstration can simulate either a discontinuous or continuous thermal inversion. To simulate the discontinuous inversion, we filled the tank with two liquids. To represent the cold, optically dense air brought to the Titanic by the Labrador Current, we added 5 gallons of a high-density sugar solution (made by thoroughly mixing 1,580 grams of common table sugar and 5 gallons of tap water). We also experimented

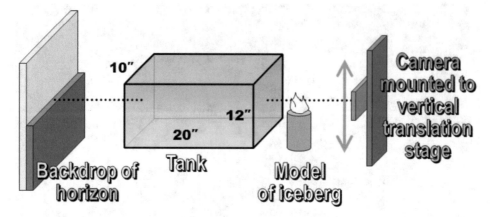

Figure 8.4: Set-up of our in-class demonstration.

with solutions of baking soda and salt, which worked well, but were cloudier than the sugar solution. Next, to represent the warmer, less optically dense air brought to the Titanic by the Gulf Stream, we simply added five gallons of tap water. To have a successful demonstration, the tap water must be added *slowly and cautiously* to the sugar solution so that the tap water forms a separate layer on top of the sugar solution. To minimize mixing of the two layers, we simply let the tap water trickle into tank using a small line of 1/2-inch aquarium hose. Obviously, some mixing of the two layers will occur, but if the tap water is added slowly, a *distinct* boundary between the liquids can be established and provides a stunning demonstration to students. The discontinuity between the layers is readily-apparent and fascinating to view. To establish a simulation of a continuously-varying thermal inversion, simply add 3,160 grams of common table sugar to 10 gallons of tap water, without stirring, and let the solution sit for two days. Over time, a continuous gradient of sugar concentration will form.

Next, we set the height of our model iceberg to match the height of the transition layer and slowly translated the camera's line-of-sight vertically across the model of the iceberg. The camera need only be translated 2–3 centimeters. The superior mirage of the horizon was easy to see as the camera's line-of-sight passed through the transition layer. Of the two types of simulations, we found that the discontinuous inversion produced the best results. Although this set-up required more preparation work, its visual effects were stark. Once our simulated atmosphere was established, the effects of looming on Titanic's lookouts were easy to understand. Moving the camera's line-of-sight across the refractive transition layer, the height of the "false horizon" was vertically displaced to a point where the iceberg was totally concealed (Figure 8.5F). No wonder the lookouts testified that the iceberg "… *was a dark mass that came through that haze and there was no white appearing until it was just close alongside the ship.*"

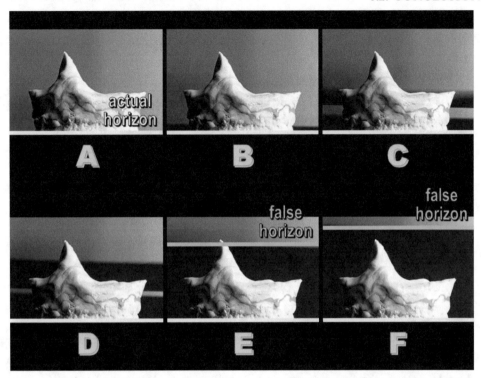

Figure 8.5: In-class demonstration. In Figure A, the actual horizon was seen at the very bottom of the iceberg. Looking at Figure A, one can imagine that a dimly lit iceberg may have been easy to spot by Titanic's lookouts. As the camera was slowly translated upward and the camera's line-of-sight passed through the transition layer, the superior mirage of the horizon began to emerge—figures C and D. In Figure E, only the very tip of the iceberg was seen rising above the false horizon while in Figure F, the iceberg appeared to be completely beneath the level of the false horizon. Looking at Figure F, one can imagine that a dimly lit iceberg may have been completely concealed to Titanic's lookouts.

8.2   CONCLUSIONS

Atmospheric refractive phenomena, poorly equipped lookouts, and the absence of a coordinated ice patrol may have caused the R.M.S. Titanic to strike a concealed iceberg on April 14, 1912. Short-sighted designs and flaws in material may have caused the ship to sink much faster than engineers ever envisioned. Finally, a false sense of invincibility, poorly conceived rescue operations, and outdated passenger liner regulations may have led to the unnecessary deaths of over fifteen hundred passengers and crew. Although not providing any definitive answers, the atmospheric refractive phenomena discussed in this paper offer some intriguing clues as to why Titanic's lookouts failed to see the on-coming iceberg on that fateful night and lend themselves to a forensics-based re-examination of the tragedy that is well-suited for the introductory physics classroom.

8.3   REFERENCES

[1] G. A. DiLisi, The R.M.S. Titanic: A case study in thermal inversion and atmospheric refraction phenomena, submitted to *Phys. Teach.*, April 2020. 119

[2] M. Grey, Photograph believed to show "Titanic Iceberg" up for auction, *CNN*, October 17, 2015. https://www.cnn.com/2015/10/17/europe/titanic-iceberg-picture-photo/index.html on April 4, 2020. 120

[3] Photograph courtesy of Wikimedia. Public Domain. 120

[4] Officers in Titanic's bridge may have been the first to see the iceberg as Fleet noted that the ship seemed to already be turning when he phoned the bridge. 120

[5] T. Maltin, *Titanic: A Very Deceiving Night*, Amazon E-books—Kindle Edition, March 19, 2012. 120, 122, 123

[6] R. Gannon, What really sank the Titanic, *Pop. Sci.*, 246:49–55, February 1995. 121

[7] Division of Work and Industry, Transportation Collections, National Museum of American History, Smithsonian Institution, The Titanic, Smithsonian Institution, Washington, DC, May 1997. https://www.si.edu/spotlight/titanic on April 4, 2020. 121

[8] Debate continues as to why the lookouts were not equipped with binoculars. Fleet and Lee testified that company policy did not require lookouts to use binoculars. Also, rumors persist that an employee, who left the Titanic before its maiden voyage, forgot to turn over the key to the locker where the binoculars were kept in storage. According to another version, the employee took the binoculars with him when he left the ship, as they were his personal set of binoculars. 121

[9] K. Ravilious, Weatherwatch: Did warm weather cause the Titanic disaster?, The Guardian, April 27, 2014. https://www.theguardian.com/news/2014/apr/27/weatherwatch-icebergs-greenland-titanic, on March 31, 2020. 122

[10] The refractive phenomena discussed in this paper may also have been the reason that the crew of the S.S. Californian did not render assistance to the Titanic. These effects may have caused the crew of the S.S. Californian to misinterpret the size of, and distance to, a nearby ship that was, in fact, Titanic. 124

CHAPTER 9

Making Local Connections

Last, case studies offer teachers and students the opportunity to explore and re-examine local events

Not all case studies have to involve historical events from the national or international level. Re-examining local events, or events that transpire at school, can bring content directly into students' lives. Thus, case studies personalize the curriculum while emphasizing how content can be used to treat a wide range of events. For this topic, we revisited a well-known controversial football play between two Cleveland-area high school football powerhouses. The play is legendary in Cleveland lore. A video analysis of the play opens up the historical debate to, *"You make the call!"* and asks students to determine if the referee indeed made the correct ruling [1]. The treatment boils down to the vector analysis of the projection of a 3-dimensional event onto a 2-dimensional surface. Specifically, the analysis examines the controversial ruling and calculates the observational error created by the camera's projection of that play onto the football field. The article graced the cover of *The Physics Teacher* when it first appeared in print (see Figure 9.1).

9.1 MONDAY NIGHT FOOTBALL–PHYSICS DECIDES CONTROVERSIAL CALL: A CASE STUDY IN OBSERVATIONAL ERRORS

Gregory A. DiLisi, *John Carroll University, University Heights, Ohio*
Richard A. Rarick, *Cleveland State University, Cleveland, Ohio*

9.1.1 INTRODUCTION

Physicists and physics educators are members of a strange breed—or so we're told [1]. Mention the word "physics" at a party, and you'll likely be eating alone. Social reactions to physicists commonly fall into one of three categories. I'll kindly label the first and most popular social response as the *"What Is That?"* reaction. I recently went to get a haircut and after some polite small talk, the barber could resist his curiosity no longer. "So what do you do for a living, pal?" he nonchalantly asked. "I teach physics," I said with my now customary trepidation. "Oh really," he said, "Can you take a look at my knee?" After I informed him that *physicists* are not *physicians*, and that we deal with the interactions of matter with the physical world, his interest in my job was suddenly replaced by talk of his latest golf game.

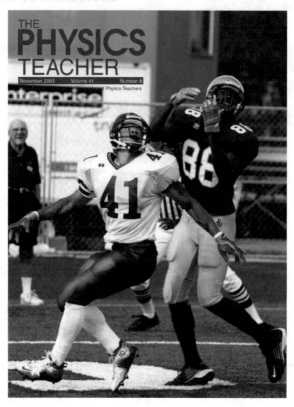

Figure 9.1: A controversial play during a football game. The projection of a 3-dimensional event (the play) is viewed on a 2-dimensional surface (the television screen). This chapter examines the relationship between the "apparent location" of the ball on the 2-dimensioanl surface with the "actual location" of the ball in 3 dimensions. Photograph courtesy of the John Carroll University Athletics Department and the American Association of Physics Teachers.

The second social reaction will be filed under *"H"* for *"Hatred,"* *"Horrified,"* or *"Hostile;"* all three equally apply. More descriptively, this reaction is labeled with the *"I Always Hated That"* moniker and is generally employed by individuals in the professional ranks. While my doctor was examining my knee after a recent arthroscopy, he chatted with me by describing how poorly he did in his undergraduate physics classes. In fact, he informed me, he could remember the names of all of his physics professors, how they dressed, their every idiosyncrasy, and how the physical science section of the MCAT exam almost prevented him from attending medical school. Recollecting his high school and undergraduate physics classes, he seemed to be having an out-of-body experience. I kept wondering if his twisting of my knee was some latent passive-aggressive response to his introductory physics ordeals.

The last social classification of physicist is courtesy of my big brother—it's affectionately called the *"That's Useless"* tag. My brother, the family lawyer, is constantly deluged with calls from not-so-immediate family members seeking help with wills, divorce papers, traffic violations, you name it. Of course, he's expected to supply his services *gratis*. One day, while lamenting his career choices, he jealously looked at me and said: "You don't know how lucky you are that nobody needs a physicist."

Well, for all of you physicists and physics educators, I'm here to tell you a story of glorious, albeit brief, self-fulfillment—somebody actually needed a physicist and I was there to answer the call. Before a departmental meeting, one of my colleagues was describing his part-time job as a referee/official for local high school football games. He had the room spellbound as he described his involvement in a recent, hotly contested, playoff game controversy. According to the official, a quarterback had completed a spectacular pass that would have won the game for his team. However, the official, claiming he possessed "the optimum viewing position," ruled that the quarterback's airborne knee (about 1.5 feet off the ground) was *beyond* the line of scrimmage when the ball was released, thus negating the play. In referee jargon, the play was ruled "an illegal forward pass" [2]. A videotape, analyzed after the game, appeared to show that the referee was in error; namely, that the quarterback's knee was instead *behind* the line of scrimmage when the ball was released (see Figures 9.2 and 9.3). The referee was adamant that he had called the play correctly and that the videotape was only showing, in his words, "the projection of the quarterback's knee onto a 2-dimensional surface." That kind of talk gets a physicist's blood pumping! He maintained that the camera had recorded the passer's knee-location against the background of the playing field, giving the illusion that he had made the wrong call. "I still have the videotape. If only someone could scientifically analyze the play for me," was his final plea.

I have constructed a general vector-based response to the problem. Below is my solution that I have since incorporated into my introductory undergraduate physics courses in Classical Mechanics. I think this problem is an excellent exercise for students to practice 3-dimensional vector analysis. Such exercises will serve students well in visualizing and analyzing other 3-dimensional problems and certainly evoke a positive response from the student sports enthusiasts. The analysis is suited for both the secondary and undergraduate levels. Finally, since the National Football League has embraced the use of video-replay as a method of correcting inaccurate officiating during a game, the analysis of the inconsistencies attributable to imaging technology is warranted.

9.1.2   STATEMENT OF THE PROBLEM

The dimensions, gridlines, and markings of a high school football field are important to the statement of the problem and to the determination of various coordinates needed to analyze the controversial call. In length, the high school football field is 100 yards with tick-marks every yard and gridlines every 10 yards. An additional 10-yard endzone exists at both ends of the field.

Figure 9.2: Video capture: At the start of the play, the ball lies in the xy-plane. Its x-coordinate is 10.5 yards (it lies just past the 10 yard marker). Its y-coordinate is 25 yards (the mid-width of the football field is approximately 25 yards). This location defines the line of scrimmage and the plane of the neutral zone. The quarterback cannot throw a legal pass if his knee or toe crosses the 10.5 yardline. Video capture courtesy of James Lanese.

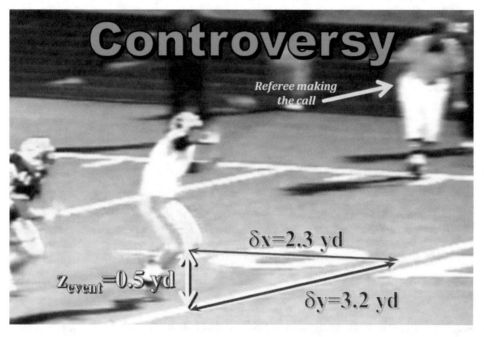

Figure 9.3: Video capture: The quarterback throws the ball and the referee rules that the player's knee had crossed the 10.5 yard-line. The apparent x-coordinate of the quarterback's knee is 8.5 yard. The apparent y-coordinate is 43 yard (he is 10 yards from the far sideline). We estimate the knee to be 0.5 yards along the z-axis. Finally, the approximate values of δx, δy, and z_{event} are indicated. Note that δx and δy lie in the plane of the football field while z_{event} is perpendicular to the plane of the football field. Video capture courtesy of James Lanese.

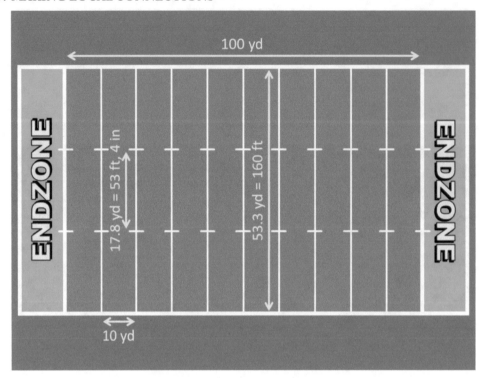

Figure 9.4: The dimensions and markings of a high school football field.

In width, the high school football field is 53.3 yards (160 feet) with hash-marks located 17.8 yards (53 feet, 4 inches) from both sidelines. A rough schematic is shown in Figure 9.4 .

To analyze the controversial call, see Figure 9.5 in which the plane of the football field defines the xy-plane. Coordinates on the x-axis are measured with respect to the sideline yardage-markers, coordinates on the y-axis are measured with respect to the hash-marks across the playing field, and coordinates on the z-axis correspond to the height above the football field (these coordinates must be estimated). A camera is positioned at a known location $(x_{camera}, y_{camera}, z_{camera})$. The camera captures an event: the location of the quarterback's knee at the instant he releases the ball. The location of this event is unknown and will be designated by the unknown-coordinates $(x_{event}, y_{event}, z_{event})$. However, from the camera's point of view, the event appears to be projected onto the football field, and its apparent location is denoted by $(x_{apparent}, y_{apparent}, z_{apparent})$. This projected location can be determined from the television screen as the videotape is played. In other words, a line passing from the camera location through the event location intersects the playing field at $(x_{apparent}, y_{apparent}, z_{apparent})$. Note that $z_{apparent} = 0$ because the location of the apparent event lies in the plane of the football field. Thus, given the

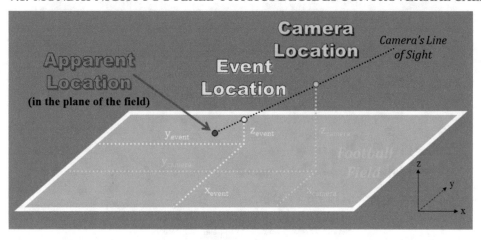

Figure 9.5: Depiction of the physics problem. Given: $(x_{\text{camera}}, y_{\text{camera}}, z_{\text{camera}})$ and $(x_{\text{apparent}}, y_{\text{apparent}}, 0)$. Determine: $(x_{\text{event}}, y_{\text{event}}, z_{\text{event}})$.

apparent coordinates of the event and the camera's coordinates, the problem is to determine the true coordinates of the event.

9.1.3   SOLUTION

From introductory vector analysis courses, a point (x, y, z) lies on the line passing through the point (x_0, y_0, z_0) and parallel to the vector $\mathbf{V} = a\mathbf{i} + b\mathbf{j} + c\mathbf{k}$, if and only if its coordinates satisfy the parametric equations [3]:

$$
\begin{aligned}
x &= x_0 + at, \\
y &= y_0 + bt, \\
z &= z_0 + ct,
\end{aligned}
\tag{9.1}
$$

where t is some scalar. In order to analyze the controversial football call, we are seeking $(x_{\text{event}}, y_{\text{event}}, z_{\text{event}})$, so we use these coordinates for (x, y, z) in Eq. (9.1). Either of the points $(x_{\text{camera}}, y_{\text{camera}}, z_{\text{camera}})$ or $(x_{\text{apparent}}, y_{\text{apparent}}, z_{\text{apparent}})$ may be used for (x_0, y_0, z_0) in Eq. (9.1), but for convenience, we choose $(x_{\text{apparent}}, y_{\text{apparent}}, z_{\text{apparent}})$ because the z-coordinate is 0 (Recall that the apparent location of the event is the projection of the event onto the xy-plane, thus $z_{\text{apparent}} = 0$). We choose $\mathbf{V}$ to be the vector, the tail of which is $(x_{\text{camera}}, y_{\text{camera}}, z_{\text{camera}})$ and the tip of which is $(x_{\text{apparent}}, y_{\text{apparent}}, z_{\text{apparent}})$. Note that either point can be used for the tip since, for vectors, $-\mathbf{V}$ is anti-parallel to $\mathbf{V}$. The components of $\mathbf{V}$ are then given as the difference in coordinates between the tip and tail of $\mathbf{V}$:

$$a = x_{apparent} - x_{camera},$$
$$b = y_{apparent} - y_{camera},$$
$$c = z_{apparent} - z_{camera}. \tag{9.2}$$

So

$$\mathbf{V} = a\mathbf{i} + b\mathbf{j} + c\mathbf{k}$$
$$= \left(x_{apparent} - x_{camera}\right)\mathbf{i} + \left(y_{apparent} - y_{camera}\right)\mathbf{j} + \left(z_{apparent} - z_{camera}\right)\mathbf{k}, \tag{9.3}$$

and our parametric equations in Eq. (9.1) become:

$$x_{event} = x_{apparent} + \left(x_{apparent} - x_{camera}\right)t,$$
$$y_{event} = y_{apparent} + \left(y_{apparent} - y_{camera}\right)t,$$
$$z_{event} = z_{apparent} + \left(z_{apparent} - z_{camera}\right)t. \tag{9.4}$$

We are primarily interested in x_{event}, the true location of the player's knee along the direction of the sidelines, (i.e., along the x-axis). The key to solving the problem is to substitute $z_{apparent} = 0$ in the third equation in Eq. (9.4), solve for t, and then eliminate t in the first and second equations:

$$z_{event} = -z_{camera}\, t,$$
$$t = -\frac{z_{event}}{z_{camera}}, \tag{9.5}$$

$$x_{event} = x_{apparent} + \left(x_{camera} - x_{apparent}\right)\left(\frac{z_{event}}{z_{camera}}\right), \tag{9.6}$$

$$y_{event} = y_{apparent} + \left(y_{camera} - y_{apparent}\right)\left(\frac{z_{event}}{z_{camera}}\right), \tag{9.7}$$

Note that Eqs. (9.6) and (9.7) are valid only if $z_{camera} \neq 0$, that is, only if the camera is situated above the playing field (as is the case).

Thus, given the location of the camera $(x_{camera}, y_{camera}, z_{camera})$, the apparent location of the event as seen by the camera $(x_{apparent}, y_{apparent}, z_{apparent})$, and a reasonable estimate for z_{event}, Eqs. (9.6) and (9.7) can be used to estimate the "true" location of the event $(x_{event}, y_{event}, z_{event})$.

Using Eqs. (9.6) and (9.7), the amount of observational error created by the camera's projection of the event onto a 2-dimensional surface can be estimated. Some special cases are useful to note: (i) if $z_{event} = 0$ or (ii) if $x_{camera} = x_{apparent}$, then the apparent and true coordinates of the event are the same; and (iii) if $x_{camera} > x_{apparent}$, then $x_{event} > x_{apparent}$. Since we know that the quarterback's knee was above the field and that the camera was downfield from the apparent event, (i.e., $x_{camera} > x_{apparent}$), we know that case (iii) applies to our controversial call and therefore $x_{event} > x_{apparent}$. Thus, the location of the true event is actually beyond the apparent location—that is, the quarterback's knee was beyond the line of scrimmage when he released the ball. This is what ultimately corroborates the referee's call.

9.1.4   RESULTS AND SUGGESTIONS

Some interesting effects emerge from this simple vector analysis. The observational errors induced by the relative camera location can be expressed in the form of $(x_{event} - x_{apparent})$ and $(y_{event} - y_{apparent})$, i.e., the difference in the x and y coordinates of the true and apparent events. If we define the x-observation error δx to be $(x_{event} - x_{apparent})$ and the y-observation error δy to be $(y_{event} - y_{apparent})$, then using Eqs. (9.6) and (9.7), the observation errors can be rewritten as:

$$\delta x = \left(\frac{z_{event}}{z_{camera}}\right) \cdot \Delta x, \tag{9.8}$$

$$\delta y = \left(\frac{z_{event}}{z_{camera}}\right) \cdot \Delta y, \tag{9.9}$$

where

$$\Delta x = (x_{camera} - x_{apparent}) \quad \text{and} \quad \Delta y = (y_{camera} - y_{apparent}). \tag{9.10}$$

Clearly, the only factors determining the values of the observation errors are the x- and y-coordinate differences between the camera and apparent location of the event, and the height (z-coordinate) of the actual event. Note that both the x and y observation errors vanish if Δx and Δy vanish. Thus, the observation errors on the television screen caused by the camera's inherent projection of a 3-dimensional event onto the 2-dimension background surface of the football field can in fact be minimized. Observation errors along the length of the football field, (i.e., from endzone-to-endzone) and observation errors along the width of the football field, (i.e., from sideline-to-sideline) can be minimized by decreasing the respective distance between camera and apparent event location. For instance, to minimize the observation error **from endzone-to-endzone**, the distance between the camera and apparent event must be minimized **from endzone-to-endzone**. Additionally, placing the camera as high as possible minimizes all observation errors occurring in the xy-plane.

9.1.5   AND WE LIVED HAPPILY EVER-AFTER

Finally, let's analyze the specific play in which my referee-colleague was involved (see *Video Captures* in Figures 9.2 and 9.3). We have digitized the play to illustrate the measurements that were used. First, we need to determine the z-coordinate of the "event," i.e., the height of the quarterback's knee at the instant the ball was released. We can estimate that at the instant the ball left the quarterback's hand, his knee was approximately 1.5 feet in the air—thus $z_{event} = 0.5$ yd. Next, we need to establish the coordinates of the apparent event; using the digitized videoclip, we estimate that $x_{apparent} = 8.5$ yd and $y_{apparent} = 43$ yd. Finally, we need to estimate the coordinates of the camera: the camera was located 5 yards beyond midfield—thus $x_{camera} = 55$ yd, about 20 yards deep into the grandstands—thus $y_{camera} = -20$ yd, and approximately 10 yards high in the grandstands—thus $z_{camera} = 10$ yd.

The important numbers to keep in mind are the *differences* in the camera coordinates and the apparent coordinates. Along the *x-axis* (end-zone to end-zone), a 46.5 yard difference exists ($\Delta x = 46.5$ yd) and along the *y-axis* (sideline to sideline), a 63 yard difference exists ($\Delta y = -63$ yd). These numbers ultimately determine the δx and δy observation errors. The results are:

Along x-axis—from Eq. (9.8)**:**

$$\delta x = \left(\frac{0.5}{10}\right) \cdot (55 - 8.5) = 2.3 \text{ yd, and}$$

Along y-axis—from Eq. (9.9)**:**

$$\delta y = \left(\frac{0.5}{10}\right) \cdot (-20 - 43) = -3.2 \text{ yd.}$$

After several phone calls, the referee and I met for pizza and discussed my "solution" to his controversial call. After distilling the analysis a bit, my new pal seemed confident that the 2.3 yard observation error along the x-axis (from endzone-to-endzone) was enough to validate his call. He even took the video and my analysis to the Cleveland Football Official Association to further corroborate his call. Needless to say, I made a new friend, finally put my physics education to some practical use, and got a free pizza to boot. If only my lawyer brother could have been there!

9.2   GENERAL VALUES

In the particular play just discussed, the referee was most concerned about the observation error along the x-axis, δx. Since δx and δy have the same form, we have plotted some generic values of only the x-observation error (measured in yards) on the field. We have plotted (see Figure 9.6) δx vs. $x_{apparent}$ for a family of z_{event} values. Assume a typical location for the camera to be at the 50 yard-line, 20 yards deep into the grandstands, and 10 yards high in the grandstands.

Also of interest, we plotted the total apparent observation error, calculated as $\sqrt{\delta x^2 + \delta y^2}$, for $z_{event} = 1$ yd at all points on the football field (see Figure 9.7). The severity of the total observation error is indicated by the grayscale. For instance, any apparent event occurring between the 20 and 30 yardlines, and within the first hash-mark, has a total observation error of roughly 4

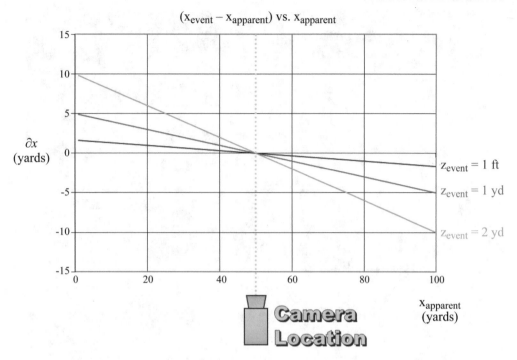

Figure 9.6: "Displacement" vs. "Apparent Location of Event."

yards. With this plot as a reference guide and videotape replay, referees and television commentators could quickly and quantitatively gauge the apparent distortions occurring on the television screens. We envision a physicist-friendly-future in which a commentator may say (using Figure 9.7): "That play occurred at the 25 yardline and inside the first hash-mark. Even though the quarterback looks to be out-of-bounds, our staff physicists tell us that there may be a 4 yard distortion on your television screens. Who says physicists aren't useful?"

9.3   ACKNOWLEDGMENTS

Special thanks to Jim Lanese, Visiting Assistant Professor of Education, John Carroll University. His scientific curiosity and job as high school football referee provided the incentive for this paper. Also, special thanks to Colleen Winters and Tom Philips for assisting with the field measurements and digitizing the video clips.

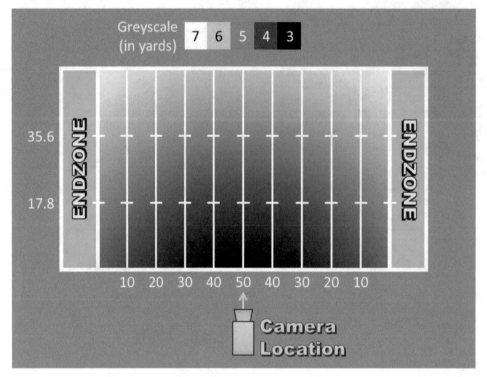

Figure 9.7: Total Displacement (for $z_{event} = 1$ yd) at all points on a high school football field.

9.4   REFERENCES

[1] G. A. DiLisi and R. A. Rarick, Monday night football: Physics decides controversial call, *Phys. Teach.*, 41:454–459, November 2003. DOI: 10.1119/1.1625203. 129

[2] The 2001 National Federation of State High School Associations Football Rulebook states: The passer commits an illegal forward pass when he has one foot beyond the plane of the neutral zone (defined by the line of scrimmage) when he releases the ball on a forward pass. The pass is illegal. An illegal forward pass is part of a running play with the end of the run being the spot from which the pass is thrown. 131

[3] P. V. O'Neil, *Advanced Engineering Mathematics*, page 393, Wadsworth Publishing Co., Belmont, CA, 1983. 135

CHAPTER 10

That's a Wrap!

**You might be wondering how case studies are received by teachers.
Fear not ... so were we.**

The articles and activities from the previous eight chapters, along with background information, additional readings, and classroom presentations, were compiled into a set of classroom modules for pre-service teachers (PSTs). These PST were senior-level undergraduate students completing their student-teaching experiences (see Figure 10.1). The PST were placed in a wide variety of academic institutions, in diverse socio-economic settings, and taught grades pre-K through 12. The format of each module alternated between seminar-style discussions and laboratory activities. To start, we asked PSTs to read the one of the chapters before class. Then, during the seminar portion of class, we reviewed the background information and used excerpts from eyewitness accounts, newsreel footage, and audio clips to authentically portray the historical event under consideration. PSTs were then divided into teams of four and moved to the laboratory where, in the spirit of the popular show *"MythBusters,"* they tested leading (sometimes competing) theories posited by scientists and historians to explain why the event occurred. Each team performed the laboratory activity described in this book, as opposed to watching a single whole-class demonstration. The small team structure was crucial to the success of our modules since an unplanned benefit ensued from this socialization—an esprit de corps emerged among members of each team while friendly competitions pervaded the laboratory to see what groups could solve each session's historical debate. The drive to resolve each module's historical debate was palpable. Teams concluded the laboratory portion of each module by using problem-solving techniques to understand and interpret the underlying physics of the event. Finally, PSTs returned to the seminar room where we discussed results and polled workgroups in an effort to reach consensus in resolving the case study's historical debate.

Figure 10.1: Pre-service teachers provided feedback on the use of case studies as a pedagogical approach to teaching STEM, especially physics.

10.1   A CASE STUDIES APPROACH TO TEACHING INTRODUCTORY PHYSICS

Gregory A. DiLisi, *John Carroll University, University Heights, Ohio*
Richard A. Rarick, *Cleveland State University, Cleveland, Ohio*
Alison Chaney, *John Carroll University, University Heights, Ohio*
Stella McLean, *John Carroll University, University Heights, Ohio*

10.2   REACTIONS TO CASE STUDIES

Eighteen pre-service teachers reflected on how they might use a case study pedagogy to plan and implement their own instruction [1]. Overall, the PSTs welcomed the opportunity to incorporate case studies into their repertoire of teaching pedagogies. According to the PSTs, the approach is an effective way for learners to contextualize physics in real-world situations that are more engaging, relevant, and exploratory than what are commonly found in more traditional teaching pedagogies. Of universal appeal to the PSTs was the open-ended nature of case studies, (i.e., that each controversy had no correct *"textbook"* solution). For example, one PST commented: *"My students will enjoy being 'detectives' and 'MythBusters,' trying to unravel why certain events happened. I know I did when I was trying to figure out the case studies we examined. I really wanted to solve each of the mysteries we explored."* Another PST commented that the case studies were so interesting and puzzling that she, *"couldn't wait to get into lab and figure out an explanation of what might have happened."*

Feedback from PSTs reveals four specific themes regarding their perceptions of the strengths associated with using a case study-based pedagogy.

- To our surprise, half of the PSTs identified our previous articles and publications in peer-reviewed journals as a powerful motivating influence for learning content. PSTs commented that seeing their instructors engaged in sustained research into these events was a strong motivation to learn the material themselves. One PST commented: *"I was excited to see my professor so engaged and personally interested in the case studies."* Another PST remarked: *"Seeing our professors doing research outside of the classroom allowed me to view them as researchers who are passionate about their work. These topics are more than just abstract concepts to them. I enjoyed seeing my professors personally invested in the case studies. This motivated me to learn more about the topics myself."*

- Next, discussing the specifics of each case study affirmed our notion that the modules were raising the historical awareness of our PSTs. Prior to class, not one PST was aware of the *Apollo I* fire, the *Hindenburg* disaster, nor the sinking of the *S. S. Edmund Fitzgerald*. These events were regarded as *"occurring in the distant past,"* (defined by one PST as any event occurring before 1990) yet were made more relevant to PSTs' lives because of the background information provided. For example, learning about the lives of the three astronauts aboard *Apollo I* made the events more personal and, as one PST noted, *"narrowed the gap between 1967 and today."* Likewise, one PST remarked about the *Hindenburg* disaster: *"Seeing the newsreel footage and hearing actual audio clips were intense experiences. I realize I have to pay more attention to events that happened outside of my own lifetime. I now understand the significance of the phrase, 'Oh, the humanity!'"* (In class, we described Herbert Morrison's eye-witness account of the *Hindenburg* disaster and phrase *"Oh, the humanity!"* as legendary audio history.)

- Next, as much as PSTs appreciated learning about events from the *"distant past,"* they all agreed that analyzing local events, as well as events occurring during their lifetimes, was a strength. Re-examining a well-known local football controversy and learning about local crewmen who perished aboard the *Fitzgerald* certainly *"brought events closer to campus ... and home."* Events like *"Deflategate"* and the Indian Ocean Tsunami sparked many discussions among PSTs about where they were when these events occurred and what they remembered about each. The take-home message here is that assembling a mix of case studies from local, national, and international events, as well as the recent and distant past, serves as a powerful catalog of situations that appeals to a wide range of students' interests.

- Finally, PSTs identified the progression of case studies, along with the ability of the instructors to transition from one study to the next, as strengths. Using a module with a local event hooked PSTs into the pedagogy. Re-examining back-to-back case studies involving the propagation of flames, followed by back-to-back case studies involving simulations, were perceived as careful course-planning on the part of the instructors. The module on *"Deflategate"* and the local football controversy was seen as bringing the experience full-circle and provided a sense of connection for the PSTs. *"I liked how each topic seemed to be a spin-off of the previous topic. This was a great way to reinforce the previous week's materials and allowed me to make connections between each topic."*

Feedback also reveals that the PSTs perceived one major limitation with using case studies in their classrooms—time. Eight respondents commented that although they enjoyed the historical nature of case studies, the pedagogy was perhaps an inefficient use of time and would prevent them from covering necessary content in their classrooms. They felt case studies were simply too time-consuming to be used on a regular basis. These respondents commented that case studies are *"good for depth, but not for breadth,"* adding that, *"the advantage of case studies is that they allow teachers to dig deeply into topics … but unfortunately, less topics get covered."* To these PSTs, traditional pedagogies like direct instruction or backward-design provided more effective means of delivering classroom instruction. One PST noted: *"Maybe a case study could be used once or twice a month to add some context to certain topics, but overall, I have to cover more material in class than a case study-approach would allow."* Unfortunately, the realities of high stakes testing and pressures from districts and administrators was never far from the minds of the PST: *"We simply have to get through a fixed curriculum so that are students have covered the necessary material when testing dates arrive."*

Finally, PSTs unanimously appreciated investigating case studies firsthand and encouraged other teacher preparation programs to institutionalize the pedagogy as part of their methods classes or student-teaching experiences. Several PSTs mentioned that whether or not they see themselves using a case studies-approach to teaching physics or STEM, they appreciated learning about it and, more importantly, experiencing it. *"There are pros and cons to every pedagogical approach and it's up to the teachers to decide what works for them, their students, and their classrooms. Adding the case study-approach to my arsenal of delivery methods gives me another choice of how I can teach STEM to my students."*

10.3 REFERENCES

[1] G. A. DiLisi, A. Chaney, S. McLean, and R. A. Rarick, A case studies approach to teaching introductory physics, *Phys. Teach.*, 58:156–159, March 2020. DOI: 10.1119/1.5145402. 142

Authors' Biographies

GREGORY A. DILISI

Gregory A. DiLisi earned his Bachelor of Science degree, with distinction, from Cornell University in Applied and Engineering Physics. He then earned his Master of Science and Doctor of Philosophy degrees in Condensed Matter Physics from Case Western Reserve University. Since then, he has taught a wide range of physics courses at the high school, undergraduate, and graduate levels. He is currently at John Carroll University, where he has held appointments in two departments — physics and education. As a faculty member, he developed over 16 courses on topics including: computational physics, experimental physics, instructional technology, interdisciplinary science, physics for engineers, problem-solving, science and society, and science methods. As an experimental physicist, he specialized in liquid crystals and complex fluids with publications appearing in peer-reviewed journals such as: *Journal de Physique II*, *Liquid Crystals*, *Microgravity Science and Technology*, and *Physical Review A*. His research focused on the viscoelastic properties and surface interactions of oligomeric liquid crystals as well as the stability of liquid bridges as they shift from micro- to hyper-gravity environments. In the area of science education, his research initially focused on developing problem-solving strategies and team-building skills in undergraduate engineering and science students. However, his current research focuses on using case studies as a pedagogical approach to teaching physics. In these areas, he has publications appearing in peer-reviewed journals such as: *The Journal of College Science Teaching*, *The Journal of STEM Education: Innovation and Research*, and *The Physics Teacher*. He has been the Principal Investigator of externally sponsored research through several grants from agencies such as: The American Association of Physics Teachers, The National Aeronautics and Space Administration, and The National Science Foundation. He was chosen to be the Ohio Educator Fellow for both of NASA's *Stardust* and *Cassini* space probes and serves as a consultant to numerous educational outreach initiatives. He has authored over 40 peer-reviewed journal articles and is an international speaker, having presented at numerous scientific and educational conferences of various professional societies.

RICHARD A. RARICK

Richard A. Rarick received his B.S. from Cleveland State University in electrical engineering and his M.S. from Cleveland State University in applied mathematics. After working in the private sector as an engineer in the fields of digital signal processing and control theory, he now is a member of the faculty in the Department of Electrical Engineering and Computer Science at Cleveland State University specializing in electronics, control theory, electro-mechanical energy conversion, and embedded systems.

Printed in the United States
by Baker & Taylor Publisher Services